Holistic Approach to Ecosystem-Based Fisheries Management

Francisco Arreguín-Sánchez

Holistic Approach to Ecosystem-Based Fisheries Management

Linking Biological Hierarchies for Sustainable Fishing

Springer

Francisco Arreguín-Sánchez
Instituto Politécnico Nacional
Centro Interdisciplinario de Ciencias Marinas
La Paz, Baja California Sur, Mexico

ISBN 978-3-030-96846-5   ISBN 978-3-030-96847-2 (eBook)
https://doi.org/10.1007/978-3-030-96847-2

This Springer imprint is published by the registered company Springer Nature Switzerland AG
The registered company address is: Gewerbestrasse 11, 6330 Cham, Switzerland

To my wife,
María de Jesús
To my daughters,
Margarita Janet
Gabriela de Jesús
To my son,
Francisco
For their love, company, and
unconditional support through the years

To the memory of
Luz María Sánchez Horta,
John A. Gulland,
David Cushing,
Luis Capurro,
John Caddy,
Daniel Lluch-Belda

Imagination should never replace reality,
but reality is built with the power of
imagination.
Drang-el M.

Foreword

When I was invited to write the foreword for this book, I felt honored and accepted the invitation immediately for the following reasons: (1) I have followed the scientific work of Dr. Francisco Arreguín-Sánchez for almost four decades, and I am impressed by his important contributions to fisheries science during this period; (2) I have, in my own studies on tropical fish stocks, also started to walk through fisheries problems as a population ecologist or classical fisheries biologist, just as Francisco did in his early career, looking at single species population dynamics, before I also became fascinated by the multispecies approaches and the holistic modeling of fished ecosystems that place the fisheries target species in the ecosystem context; (3) I wanted to read through—and learn from this new book, which condenses 40 years of fisheries research by a researcher who has not only contributed numerous scientific papers to fisheries science, but who has also formed a great number of excellent fisheries scientists in Latin America.

The concepts behind the holistic trophic modeling approach, so nicely developed in the 1980s and early 1990s by Jeffrey Polovina, Villy Christensen, Daniel Pauly, and others, and greatly enriched by the theoretical work of Robert Ulanowicz and Carl Walters, paved the way for a worldwide scientific movement towards ecosystem-based fisheries modeling and management, of which Dr. Francisco Arreguín-Sánchez is one of its greatest proponents.

The book presented here is entitled *Holistic Approach to Ecosystem-Based Fisheries Management: Linking Biological Hierarchies for Sustainable Fishing* with the key words being "holistic" and "biological hierarchies," which urge us to look at the whole but also consider the interdependencies between individuals, populations, and the ecosystem.

The book develops a coherent basis for an Ecosystem Approach to Fisheries Management (EAFM), which takes into account that (a) resources within ecosystems are interlinked among each other through nonlinear prey–predator relationships, (b) the fishery of a specific resource does not only impact the biomass pools of the fished species but also that of the interlinked species and may also directly impact the habitat, (c) changes in the environment (both stochastic ones and trends caused by global climate change) affect species performances, geographic distributions, and entire food webs. The new approach uses indicators about the state of the ecosystem to define harvest rates for individual fish resources or populations. These harvest rates correspond to the critical thresholds for the ecosystem, after which, ecosystem organization and functioning are affected, eventually losing its resilience capacity. The fishery impact is thus measured holistically, taking into account the hierarchical organization of ecosystems of individuals, populations, and the entire ecosystem.

The book is structured into 11 chapters, each preceded by an abstract. Chapter 1 provides a historical overview on the global development of (industrial) fisheries in the past century and of fisheries science from the pioneering work of Baranov, Schaefer, Beverton and Holt, Gulland, Cushing, and others in the first half of the century focusing on the population dynamics of fisheries resources under the assumption of rather constant environmental conditions and carrying capacities for the fisheries stocks—to the modern multispecies approaches for fisheries modeling and management under the consideration of varying conditions over time and space with consequences for the harvest potential for resources and ecosystems. It is shown that this has led to a paradigm shift—the Ecosystem-based Approach to Fisheries Management (EAFM), which explicitly recognizes the effect of the environment and the dynamics within and between biological hierarchies (populations and ecosystems). Chapter 2 further elaborates on the need for an Ecosystemic Approach for fisheries management, and it is argued that holistic system attributes should be used as reference indicators for applying control measures to fisheries, thus generating a "from the ecosystem to the fish stocks" management perspective. Chapter 3 provides the theoretical basis of energy flow studies in ecosystems focusing on the concepts of ascendancy, entropy, overhead, and further topological indicators used for trophic network analysis, developed by Ulanowicz and followers and elaborates on the aspects of ecosystem metabolism and the balance between supply and demand (SDB) of energy to sustain the food web. An application of the SDB concept is shown using as example the ecosystem of the central Gulf of California, where changes in the ecosystem metabolism are related with the increase in fishing intensity. Chapter 4 elaborates on the dynamic relationship between the two biological hierarchies, population and ecosystem, to better understand their dynamic interdependence in the face of the challenges of fisheries management aimed at sustainability. The concept of surplus production is presented, which is based on the balance between the primary production required to sustain consumption and for sustaining fishing along the different trophic levels. The application of this concept for the Gulf of California ecosystem is shown, where specific harvest rates can be defined for the different species of the ecosystem, being consistent with their life history. In this chapter, the concept of balanced catch is also introduced, which is based on the general idea that ecosystem harvest may be maximized if biomass from different trophic levels is being removed in proportion to the production capacity of each level. Chapter 5 provides the theoretical basis of holistic ecosystem indicators. The use of the "self-organization capacity index" is illustrated with an example from the southern Gulf of Mexico, and the "window of vitality" is described, which represents the limits of the sustainability of the ecosystem, being exemplified with the ecosystem of the Northern Gulf of California. Both the indicators, self-organization capacity and window of vitality, show that a similar condition of sustainability, estimated through holistic attributes, can be found for an ecosystem under different organization and functioning states. On the other hand, the relationship between ascendency and overhead is used to describe the concept of ecosystem health, which is exemplified, by simulation, through the loss of biomass of different species with different life histories of the ecosystem of the southern Gulf of Mexico. Chapter 6 reviews the concept of system resilience starting with Holling's adaptive cycle, which describes the performance of the system in four stages of continuous evolution: growth, conservation, collapse, and reorganization, with each of them characterized by changes in resilience. Under the ecosystem dynamics theory of Ulanowicz, the overhead ("strength in reserve"), and specifically the flow component, is an indicator of the resilience in the trophic web. Two examples are presented for illustration: the southern Gulf of Mexico that shows changes in resilience as a consequence of changes in the organization and functioning of the ecosystem due to climate change, and the central region of the Gulf of California where a threshold of resilience is reached due to a loss of biomass of the species in the ecosystem. These thresholds are represented in terms of the limit harvest rates for each species, essential information for fisheries' management. Chapter 7 reviews the concept of biological reference points applied to harvested populations. Based on these concepts, the definition and construction of ecosystem biological reference points, EBRP, using holistic attributes is discussed, on the basis of the theory of ecosystem dynamics represented by trophic networks. The performance of ecosystem indicators as a consequence of the gradual loss of biomass of some of its components (species) is described, and it is shown that CMSY overpassed those EBRP based on holistic attributes, which suggest the risk of taking MSY as target for fisheries. Chapter 8 focuses on the management of fisheries under a holistic approach and the concept of dynamic sustainability is discussed, recognizing that populations adapt to environmental changes and ecosystems continually reconfigure. With examples from top predator fisheries in the Gulf of California and shrimp fishery in the Gulf of Mexico, it is shown that overfishing emerged, since the stock management had not adapted (through a reduction in harvest rate) to the environmentally driven reduction in stock productivity. The concept of ecosystem vulnerability is also reviewed in this chapter, and systems are compared with regard to their differences in vulnerabilities. It is shown that systems are more affected if disturbances originate at low trophic levels, and ecosystems with greater organization, such as coral reefs, are more vulnerable than less organized systems, such as coastal lagoons and continental shelves.

Chapter 9 addresses adaptability as a management strategy and presents examples from the shrimp and grouper fisheries of the Campeche Bank, Mexico, that reveal that ecosystemic changes had largely negatively impacted both resources, while management had failed to adapt to these changes. Chapter 10 provides examples for ecosystem-based fisheries management. With the example of the Monterrey sardine (*Sardinops sagax*) fishery in the central Gulf of California, it could be shown that the threshold harvest rate concept led to the adaptation of the catch to the available biomass, which helped avoiding overfishing. A further example is presented with the

Mangrove community in Magdalena Bay, South Baja California. By the use of simulation experiments, it could be shown that a system threshold was reached if the mangrove cover was reduced beyond 30%. At a further loss, the aquatic ecosystem adjacent to the mangrove would lose its resilience, its capacity for reorganization. The red grouper (*Epinephelus morio*) fishery from the Campeche Bank is used again as an example for a greatly deteriorated stock that may be harvested sustainably by defining a biomass limit that allows maintaining the organization and functioning of the ecosystem and by allowing no more than 42% of the available biomass to be extracted in any given year. The pink shrimp (*Penaeus duorarum*) fishery from Campeche Bank, which supposedly had collapsed in the year 1995 due to overfishing, is a further example given. It is shown that stock size and performance were greatly affected by the prevailing climatic conditions. Given that the ecosystem had undergone continuous reconfiguration for several decades, the harvest rate that allows sustainable fishing and a sustainable ecosystem was estimated at a maximum extraction of 33% of the available biomass each year, which allows, in addition to the restoration of the population biomass extracted by fishing and the maintenance of ecosystem organization and functioning. Red octopus (*Octopus maya*) from Campeche Bank, the hawksbill turtle (*Eretmochelys imbricata*) from Campeche Bank, and the top predators in the Gulf of California are presented as further examples that show the enormous importance of changes in environmental conditions for the performance of those populations. Management of those species needs to be based on the knowledge and consideration of the environmental state of the ecosystem. While the general warming trend observed in the region favored octopus, the hawksbill turtle and top predatory fish were negatively affected. In the conclusion Chap. 11, the problem of the proposed management strategy "from the ecosystem to the individual populations" is discussed. It is noted that the reference points or performance indicators of the trophic network, when estimated from the energy flows dynamics, are not tangible and hard to understand by the fishing sector. However, the justification for the consideration of holistic reference points of the ecosystem lies precisely in the need for management under conditions of continuous changes in population dynamics and ecosystem reconfiguration over time under conditions of global climate change and man-made system alterations.

I like to congratulate Dr. Francisco Arreguín-Sánchez for providing us with an excellent foundation for a modern application of the Ecosystem Approach to Fisheries management.

ZMT, Bremen, Germany Matthias Wolff

Preface

Some time ago, in the middle of the 1980s, when I planned my doctoral research, I was speaking with John A. Gulland, I proposed to him to work on a multispecies model, with explicit interdependencies. John replied that this was useless, explaining that multispecies models, of course referring to that moment, had not proven significant additional information for resource management. It seemed contradictory because John had just published a paper where he pointed out the importance of researching more about natural mortality, which was usually assumed as constant in population models. John pointed out the importance of the predator–prey relationships as the main source of natural mortality, and the problem of the no consideration of its variability in stock assessment practice. Two lessons I learned from that answer: the first, that scientific work must have a practical sense, in this case, its application to the management of fish resources in benefit of the people who depends on a sustainable fishing; and the second, the importance of trophic interdependencies in fishery estimates, and that this was an issue yet to be developed.

Later, towards the beginning of the 1990s, I had the opportunity, since then, to work close with Daniel Pauly and Villy Christensen in the application of the new ecosystem trophic modeling platform Ecopath with Ecosim; and few years later, I meet and interact with Robert Ulanowicz who, as will be seen later, strongly influenced the development of the concepts developed in this book related to ecosystem dynamics. Along this path, there was the opportunity to address the issue of trophic relationships, and how these affects, directly and indirectly, populations within ecosystems, and in turn fisheries. At the same time, a strong demand for knowledge began to be generated that would provide more detail on the estimates of fishing mortality and available biomass. On the one hand, many of the resources were being exploited to their maximum production capacity and for their persistence, greater detail had to be provided on the effects of fishing; for example, the differential fishing mortality on breeding adults, or allowable harvest rates on juveniles, which, in turn, also need more precise estimates of natural mortality. Additionally, evidence of changes in the environment began to be observed and documented, mainly associated with climate patterns, and that these affect the production capacity; and importantly, of the target populations for fishing; until then, issues are practically not considered in the trends of changes in the abundance of fishery resources.

Currently, the scientific evidence documents (1) the effects of fishing on natural systems, (2) the effect of changes in climate patterns on species, and (3) the trophic and non-trophic interdependencies among species of an ecosystem, and evidences and theoretical knowledge increases about the behavior and dynamics of natural systems, their trophic behavior, and the needs of fisheries management. Also, at present, we can read about the so-called management based on the ecosystem, which, when practiced, it essentially refers to direct impacts of fishing on ecosystem, which could be related to incidental fishing, and on the physical impact of fishing gears on habitats. Both the aspects have different perspectives, either due to the magnitude of the biomass retained incidentally and many times discarded or due to the effect on critical habitats, or on threatened species. For the different situations, specific actions have been suggested to reduce the impact, translating into controls on fishing in terms of the harvest rate. However, all of them are aimed to the direct effects, that is, considering those species or habitats that come into contact with fishing gear. In this sense, there is a growing concern for the conservation of the organization and functions of ecosystems as base for the sustainability fishing.

The book recognizes three entities of nature that have their own dynamics, each one with their own performance indicators that are interacting and interdependent with each other and that are nested ones inside the others. These entities are individuals, populations, and ecosystems. Management measures, for example, can go to the protection or improvement of individual conditions for reproduction or growth; populations, made up of individuals, measure the success of their performance in terms of their survival; and populations, by interacting with each other through predator–prey relationships, define trophic networks whose structure, organization, and operation depend on how these relationships and energy flows occur, their indicators being a reflection of these attributes. In this panorama, it is evident that if the climatic patterns change significantly, they will affect the efficiency of reproduction and growth, and this in turn to the survival of such luck the population adapts, and these changes at the same time affect the energy exchanges in the food web. But even more, at the level of trophic interactions, as an example, two populations can behave differently according to the availability of prey; that is, if there is little or a lot of prey's biomass, the way in which such biomass is distributed and made available changes. This essentially means that the relationships between species are not linear; that is, they do not change in the same proportion according to changes in abundances. In this sense, the book highlights the definition and use of ecosystem performance indicators based on holistic attributes of the food web, not from the populations. The next step that is offered as a contribution to science is to use these performance indicators as biological reference points of the ecosystem, EBRP, to define specific harvest rates of the exploited populations, that loss of biomass, in this case because of fishing. These EBRPs are analogous to those derived from population dynamics and have the same potential use; only in this case, we speak of fishing control in relation to its effect on the organization and functioning of ecosystems. Note that, in this case, EBRPs are derived from attributes of the higher hierarchy level towards the elements on the immediate lower level, populations.

The current concept of the ecosystem-based management strategy is defined here as management strategy “from the individual populations to the ecosystem,” highlighting the focus on the direct effects of fishing on target stocks and towards the ecosystem, and the approach or concept of management strategy “from the ecosystem to the individual populations” is proposed. The idea in this approach is precisely to use holistic ecosystem-based indicators to define harvest rates for individual fish resources or population. These harvest rates correspond to the critical thresholds for the ecosystem, for example, the loss of biomass level of any component of the trophic web, after which, ecosystem organization and functioning are affected, or the loss of ecosystem resilience capacity. In this context, these thresholds correspond to EBRP_limit. The problem in practice is that these types of indicators are not tangible like those derived for populations that they are ultimately based on biomass. The EBRPs are based on energy flows of the food web, an intangible aspect. In this way, its application requires a strong confidence of the fishing sector in science, which is not an easy aspect to reach since they have not yet been applied in real cases, with a couple of exceptions aimed to certification practice. The suggestion in this sense is to use them in a precautionary sense, that is, take advantage as a precautionary EBRP, which allows considering ecosystem attributes immediately and accumulating experience to generate the required trust.

Globally, the effects of climate on the abundance of populations in marine ecosystems are recognized; individuals respond in terms of their individual physiology, being more or less efficient according to their own physiological state. These individuals make up populations, and their performance is measured by their demography which, through their survival, defines their performance. It is the role of science to determine, with the greatest possible precision, the fishing mortality regarding other sources of mortality to define the state of exploitation of the resources. In this context, populations adapt when affected by changing climate patterns, through their dynamics, to the conditions that the environment defines. The intensity and behavior of the trophic interdependencies are defined from the relative abundances of the populations, in such a way that if their abundances change as a result of the effect of climatic patterns, the ecosystems will also continually reconfigure. This aspect will be reflected in the behavior of the performance indicators of the food web. In this context, EBRPs are very useful because they define attributes according to the organization of the system’s flows, but not of the particular biomass of the populations. In accordance with this, an EBRP allows, within its oneself magnitude, to maintain the dynamics of interactions among the populations, including changes in biomass. For example, if an EBRP_precautory is defined through a harvest rate for a resource of $HR = 0.40$, it means that no more than 40% of the available biomass should be extracted per unit of time but that the level of resilience can be maintained with different combinations of biomass of the ecosystem populations. If this level of resilience reflects the sustainability of the global production capacity of the ecosystem, the concept is of interest to fisheries management, even under conditions such as the current ones, where populations adapt to changes and ecosystems reconfigure. Finally, it is considered that this book represents only one more step in the contribution of scientific knowledge supporting fisheries management. It is expected not only to be criticized but to be improved and surpassed in scientific terms for its practical utility.

La Paz, Baja California Sur, Mexico Francisco Arreguín-Sánchez

Acknowledgments

The conceptualization of the elaboration of the book was developed within the context of the research project (CB-221705) supported by the Consejo Nacional de Ciencia y Tecnología, CONACYT, called "Dynamic processes through biological hierarchies: from populations to ecosystems." Although many of the theoretical aspects were already developed, the consolidation, through the application and the gain of experience, was given thanks to the support of the Agencia Mexicana de Cooperación Internacional para el Desarrollo, AMEXCID, and the Agencia Uruguaya de Cooperación Internacional, AUCI, to the project "Development of scientific capacities under international collaboration for ecosystem-based management of marine resources in Mexico and Uruguay in the face of global climate change." In this project, the central concepts associated with the ecosystem performance indicators were consolidated when applied to various ecosystems in Mexico, Uruguay, and Colombia. These agencies are greatly appreciated for their support.

Of course, I make special mention of my own institution, the Instituto Politécnico Nacional, IPN, through the Centro Interdisciplinario de Ciencias Marinas, CICIMAR, for the support, year after year, through the Secretaría de Investigación y Posgrado, SIP, as well as for the support through the Estímulos al Desempeño de los Investigadores program, EDI, and the Comisión de Operación y Fomento de Actividades Académicas, COFAA.

I also thank the colleagues and institutions of the *Ecopath with Ecosim*, *FishBase*, and *Too Big to Ignore*, Consortia, with whom over the years and in different ways we have exchanged experiences directly related to the concepts contained in this book. Finally, and in a very particular way, I thank my colleagues from the Laboratorio de Dinámica y Manejo de Ecosistemas Acuáticos, from my own institution, without whom this book could not have been developed, thank you very much. Thanks also to Dr. Pablo del Monte Luna who drew some of the surface figures of Chap. 10.

I have deliberately avoided mentioning people's names. I would always commit the unforgivable sin of unintentionally omitting someone.

Contents

Chapter 1
Introduction

The Context of Fisheries at the Present

According to the FAO (2020), global catches have remained relatively stable since the mid-1990s, varying between 80 and 90 million tons annually. However, there are issues of great scientific concern regarding the management and sustainability of living marine resources. Let us consider the concept of sustainability, defined by the United Nations (UN 1987) as actions that "meet the needs of the present without compromising the ability of future generations to meet their own needs and ensure the balance between economic growth, care for the environment and social welfare." Regarding fishery resources, sustainability, as objective, and from the perspective of biomass production capacity, was conceptualized as obtaining of the maximum sustainable yield, which assumes, through yield models, a relative stability of ecosystems and of the exploited stocks, which means a stable or unchanging carrying capacity, a concept that recognizes variability in abundance, but without a trend (Schaefer 1954, 1957). The concept was widely accepted by the scientific community, and towards the middle of the last century, it served, for years, as the most important benchmark for fisheries management. This concept dominated scientific work for decades and was quite reasonable and sufficient for the knowledge needs that the fishing industry required at that time and, in turn, for resource management.

In the last two to three decades, the scientific community has expressed its concern over the state of fishing resources, given that, at present, many are being overexploited (Worm et al. 2009; Ye et al. 2013; Pauly et al. 2003; Srinivasan et al. 2012; Cury and Pauly 2020). FAO (2020) reports that only 34.2% of the world's fisheries are overexploited, and the problem lies in the large collapses observed in several of these resources (Fig. 1.1), impacting industries, to those who are directly and indirectly dependent on catch activity and, obviously, food security. The concern is valid from the very moment that excessive fishing capacity is recognized (Bell et al. 2017), i.e., when it is globally estimated that, if the current fishing

F. Arreguín-Sánchez, *Holistic Approach to Ecosystem-Based Fisheries Management*, https://doi.org/10.1007/978-3-030-96847-2_1

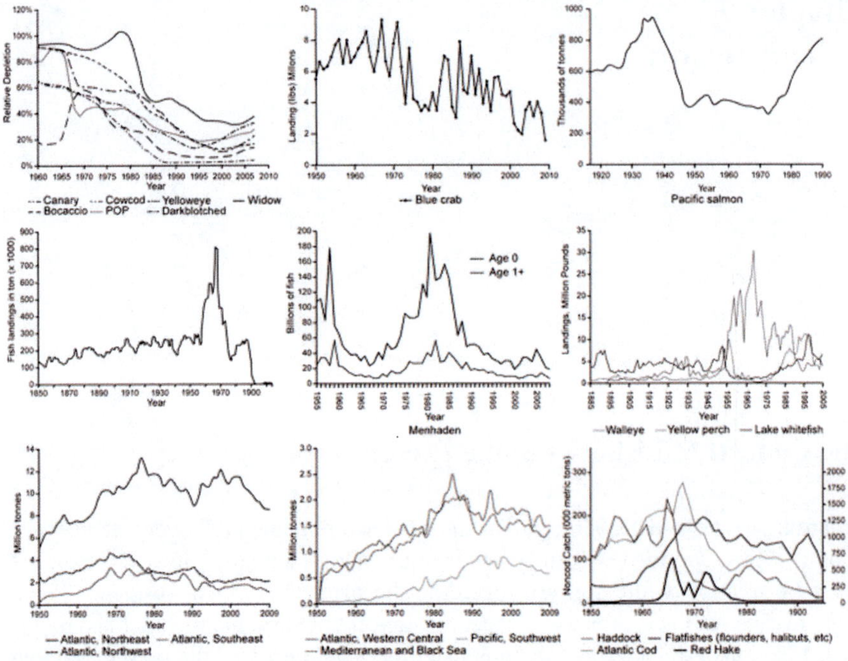

Fig. 1.1 Historical trends in the catches of different fisheries and regions of the world; for species with different life histories and that inhabit diverse ecosystems and communities. Note that in all cases, whether the trend is up or down, the inflection point is around the time of regime shift occurred, around 1980

capacity is reduced, total catch would not be affected. This, in turn, clearly indicates overcapitalization in some important sectors of the fishing industry (Cochran 2021).

If we assume that more than 59% of the resources are exploited to their maximum production capacity (FAO 2020), it follows that environmental variability, especially unfavorable conditions, could put resources at risk; further, if these unfavorable conditions are maintained, the risk of collapse increases substantially. Likewise, the spatial distribution of species abundance has changed for various resources, and in the most severe cases, there is a complete absence of some species in places where they were previously abundant (i.e., Brander 2010; Link et al. 2011; Engelhard et al. 2014; Bell et al. 2015). When speaking of exploited species, those changes are often attributed as effects of fishing even when in other cases it has also been documented as natural reasons. In any case, these assumptions and concerns rest strongly on the expectation that, by correctly applying the management measures recommended by scientists, the resources, once the conditions are favorable, will be recovered to their previous production capacity, an assumption that at present does not seem reasonable.

Concern on Climate Patterns and Fishing

In this context, the central concern of science lies in an unprecedented situation. If the trend of global sea surface temperature anomalies is taken as an indicator of environmental forcing, it is clear that today, there is an unprecedented new situation, in terms of the management and conservation of resources (Fig. 1.2). Temperature gradients, current fishing pressure on marine resources and human pressure to access food are some of the factors that, combined, make the current era an unprecedented challenge for fisheries science, for example, climate change affects the metabolism of organisms; populations are continuously adapting to these changes in different ways according to their life histories, and ecosystems are reconfigured in terms of biological interdependencies, mainly tropically, and by their habitat dependency degree. All being aspects that shape and confer a certain level of organization and cohesion to ecosystems.

On the other hand, science has also expressed great concern about the direct effects of fishing on ecosystems, such as the large volumes of bycatch (Holland 2010; Kopmoroske and Lewison 2015; Gilman et al. 2020; Kennelly and Broadhurst 2021) and their effects on the structure, organization, and function of ecosystems (Arreguín-Sánchez et al. 2002; Salcido-Guevara et al. 2012; Lotze et al. 2011); the effects of excessive exploitation of species that are considered forage for other organisms in the ecosystem (Arreguín-Sánchez et al. 2021); discards (Suuronen and Gilman 2020); the effects on habitats that in turn affect the abundance and production capacity of species and their spatial distribution (Lambert et al. 2014; Eggleton et al. 2018; Rijnsdorp et al. 2018; Porobic et al. 2019; Takeshige et al. 2021); the production capacity of the ecosystem (Ye and Carocci 2018; Link and

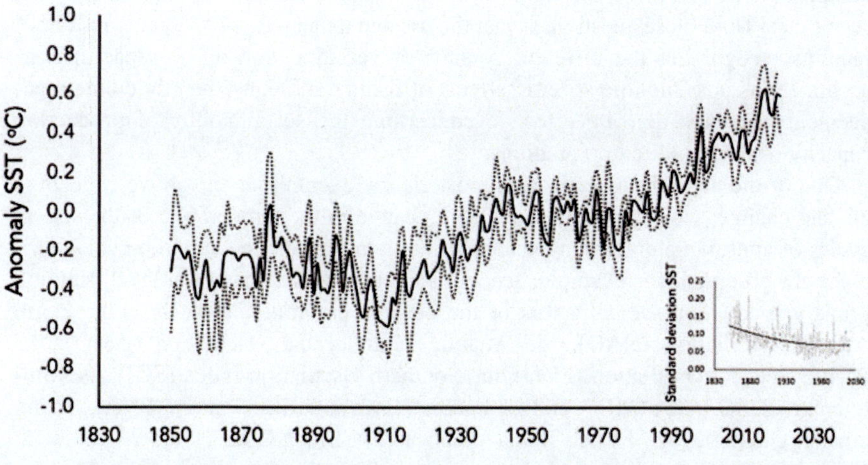

Fig. 1.2 Historical annual anomalies of the sea surface temperature showing a stable period from 1940 to 1980. The regime shift around mid-1970s shows the beginning of a gradual warming trend. Dotted lines show minimum and maximum monthly averages within each year. (Source Kennedy et al. 2011a, b)

Watson 2019; Link et al. 2020); the effects of overfishing on other non-target species, including those of commercial interest (Zhou et al. 2011; Ragheb et al. 2019); the possible undesired effects on biodiversity, including threatened species (Thrush et al. 2001; Coleman and Williams 2002; Fariñas-Franco et al. 2018; Martins-Vieira et al. 2020); and even on species or groups of species that are key to the sustainability of the ecosystem, not to mention their role in its organization and functioning (Coll et al. 2016). Similarly, much of the science concern, beyond the organization and production of natural systems, manifests in the current and potential consequences on the use of resources, and specifically, regarding the deterioration in social welfare and food security (Fabinyi et al. 2017; Stacey et al. 2018; Teh and Pauly 2018; Bell et al. 2018; Cabral et al. 2019; Loring et al. 2019).

In the current condition of environmental forcing, where populations adapt and ecosystems are reconfigured continuously, in addition to the inherent variability of the natural processes, science must generate knowledge with appropriate approaches to the new circumstances. From this perspective, the following are some of the unavoidable scientific questions: to what extent is the decrease in the abundance of a fishery resource a consequence only of fishing? How variability of climate patterns (e.g., climate change) contribute to the changes in populations' abundance? How both environment and fishing combine to drive stock abundance? or to what extent are these changes representing the adaptation of populations to the new or different environmental conditions prevailing or will prevail along a given period of time? or to what extent is the change or reconfiguration observed in the functioning and organization of an ecosystem a manifestation of the differential response of populations, and of their interdependencies, to the change in climatic patterns, or because of fishing, or both? or how can the impacts of indirect anthropogenic changes, such as global warming or incidental catch, be addressed? or adding complexity, how are the different conditions mentioned above combined? What seems clear from these questions is that the use and management of aquatic resources requires recognizing the different sources of variation that allow separating, as accurately as possible, the specific effects of fishing and consequently define management measures that are effective concerning the sustainability of production capacity of the exploited populations.

One of the natural processes that most clearly exemplifies the above concept is climate change, which is represented as a change in the temperature of the sea, in cycles of approximately seven decades. If the trends in diverse environmental variables are observed, for example, sea surface temperature, mean sea level, salinity, primary production, or indicators of the oceanic environment, such as the North Atlantic Oscillation (NAO), the Atlantic Multidecadal Oscillation (AMO), the Pacific Decadal Oscillation (PDO), the Southern Oscillation Index (SOI), the Multivariate ENSO Index (MEI), and the Pacific-North American Pattern (PNA), among others (Glantz 1992; LLuch-Belda et al. 1999; Lluch-Cota et al. 2010, 2018; Martínez-Arroyo et al. 2011; Lehodey et al. 2006; Barange et al. 2014; Brander 2007; Pepin et al. 2020), in all of them, a regime shift is observed in approximately 1980 (Fig. 1.3). This regime shift can be described as the transition between two climate patterns with opposing trends, to which populations and ecosystems respond

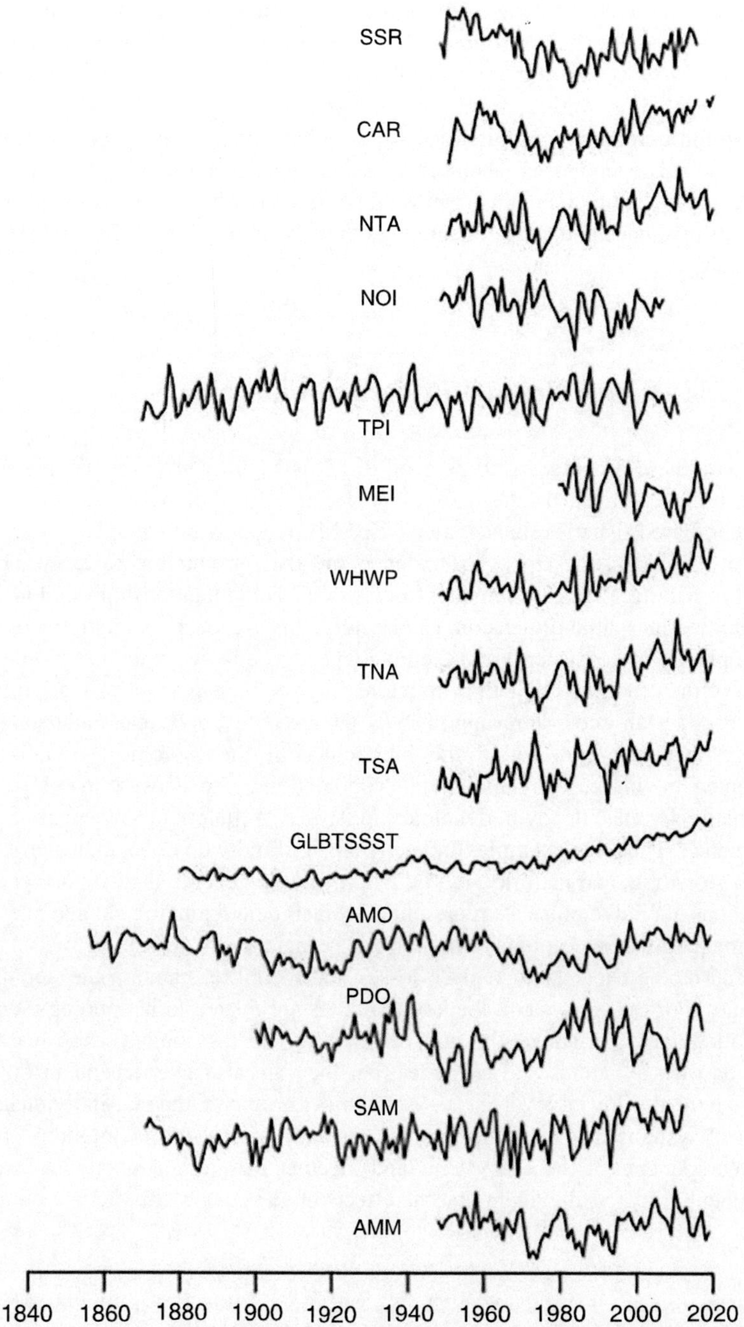

Fig. 1.3 Historical trends of several climatic and environmental variables for different regions of the world. Note that a change in trends is observed in most of the variables towards 1980. SSR, Surface Solar Radiation; CAR, Monthly Caribbean SST Index; NTA, North Tropical Atlantic Index; NOI, Northern Oscillation index; TPI, Tripol index for the Interdecadal Pacific Oscillation;

by adapting and reconfiguring itself, according to the evolution of the new conditions. In the case of fishery resources, this, in addition, manifests as changes in their production capacity and therefore in the biomass available for fishing, with these changes also being continuous. Within ecosystems, these processes of change are observed through the reconfiguration of interspecific relationships (mainly trophic) and of species' dependence on habitats; as a consequence, these processes of change are reflected in the function and organization of the ecosystem as well as in attributes such as its resilience and cohesion, among others, or in properties of interest such as biodiversity.

Biological Hierarchies and Fisheries Management

Faced with these processes, and demand of knowledge by resource use managers, science has recognized the need to adopt the principles of what has been called ecosystem-based fisheries management (EBFM), recognizing some key sources of variation that affect the stocks' abundance and their production capacity: (1) the amount of fishing, (2) environmental forcing, (3) trophic relationships, and (4) habitat including its spatial dimension. Generically, this approach explicitly recognizes the complexity of ecosystems and the interdependence between their components.

One of the scientific challenges in addressing the management of marine living resources with an ecosystem approach is the need to know and understand the dynamics between and within the hierarchies of the biological organization, represented in this case by individuals, populations, and ecosystems (Fig. 1.4). Each hierarchy has its own dynamics in terms of function, organization, and performance. Thus, for example, the metabolism of individuals regulates processes such as growth, gonad maturity, reproduction, and, in general, their robustness. For populations, their dynamics are represented by their demographics, change rates, and recruitment processes, among other factors; efficiency is measured by survival, expressed as the capacity to replace losses and therefore maintain the population continuity. For the ecosystem, the key attributes are related to its topology, energy flows, modularity, organization, and functioning. In this context, each hierarchy shows its own dynamics and attributes, but they are also interdependent on each other, where the lower level is nested within the upper; that is, individuals are biological systems that make up populations and, in turn, the populations are the living components of the ecosystems and together maintain a functional balance. Consequently, it is also evident that the effects on the ecosystem affect populations

---

Fig. 1.3 (continued) MEI, Multivariate El Niño-Southern Oscillation; WHWP, Western Hemisphere Warm Pool; TNA, Tropical Northern Atlantic index; TSA, Tropical Southern Atlantic Index; GLBTSSST, Global Average Temperature and SST Anomalies; AMO, Atlantic Multidecadal Oscillation; PDO, Pacific Decadal Oscillation; SAM, Southern Annular Mode; AMM, Atlantic Meridional Mode

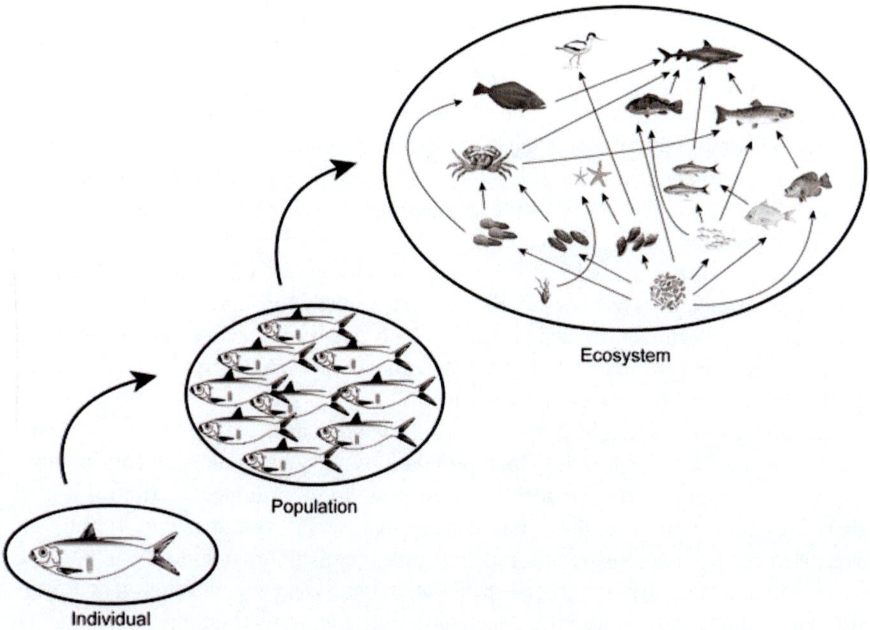

Fig. 1.4 Hierarchies of the interacting biological organization levels relevant for fisheries management, represented by individuals, populations, and ecosystems

and individuals, and it is clear that in each biological hierarchy, there will be a response to external stimulus according to the nature of its organization and functioning. In this way, it is easy to see how the knowledge and understanding of these processes are the basic elements that will ensure the sustainable use of the production capacity of resources and ecosystems, even when they show variability and a tendency to change over time as response to climate patterns. This knowledge is the scientific basis for adaptability and sustainable management.

Spatially, the effects of climate change will also manifest themselves, in a general way, in the habitats of the species and, of course, in the use that species make of it, and will be affected according to their degree of dependence. This context is highly relevant due to the multiple and varied forms of adaptation of species, which in turn will provoke diverse responses. The obvious consequence will be the need to adapt the forms of fisheries management and control to the new condition; that is, adaptability in management must reflect the adaptability of populations and ecosystems. In relation to habitat changes, the adaptability of populations to climate change varies strongly according to species' life histories; highly mobile species are able to seek more favorable conditions, as typically occurs with large pelagic organisms such as tuna (Dufour et al. 2010; Erauskin-Extramiana et al. 2018), whales (Scheinin et al. 2011; Derville et al. 2018; Meynecke et al. 2020), billfishes (Carlisle et al. 2017; Woodworth-Jefcoats et al. 2017; Erauskin-Extramiana et al. 2020), sharks (Chin et al. 2010; Hazen et al. 2012; Birkmanis et al. 2020), and turtles (Poloczanska et al. 2009; Patino-Martinez et al. 2011; del Monte-Luna et al. 2012; Almpanidou et al. 2018). In these cases, climate probably tends to have little effect on their production capacity. In contrast, if species have limited or no mobility, any negative impacts to their habitat will be reflected in their survival rates or in the inability to restore the losses, manifesting as a decline in reproductive success and, consequently, recruitment. This condition will cause decreases in species abundance and its production capacity. For example, sessile or demersal species have an affinity for specific substrates, such as bivalve molluscs, groupers, reef, and demersal fish (Arreguín-Sánchez et al. 2015, 2019; Vargas-López et al. 2021). In both the cases, species are affected in terms of their vital rates, behavior, and response capacity, translating into changes in abundance and production capacity and, eventually, reconfiguring ecosystems. If we refer specifically to fishery resources, biomass availability and fishing yields will be affected.

According to Holling and Gunderson (2002) and considering the aspects emphasized in the current scenario, in which populations and ecosystems are continuously changing according to their capacities to respond to environmental stimuli, it is of great importance to recognize that natural and social systems, considering the interest in the use and conservation of resources, constitute complex systems organized under a nested hierarchical structure, whose dynamic development occurs through nonlinear processes within each biological hierarchy (Holling 2001), and the dynamics of the interdependences between hierarchies. Figure 1.4 shows the three levels of biological hierarchy on which fishing affects: the individual level, with processes such as growth efficiency and reproductive potential; the population level, which contains the set of individuals interacting with each other and where the processes relevant to fishing have to do with the group, mainly survival, recruitment, migration, production capacity, resilience, among others; and the ecosystem level, which refers to the arrangement of populations interacting with each other and their habitats; being relevant at this level are aspects such as the topological structure of the trophic network, its organization, and capacity for self-organization, resilience, and entropy, among others.

Biological Hierarchies, the Campeche Bank as an Example

Of course, the scientific challenge is to unravel the nature of these relationships for the benefit of resource use. Perhaps a hypothesis by way of an example could be developed on the effect of climate change involving the three biological hierarchies, using independent evidence, in the region of the Bank of Campeche in the southern Gulf of Mexico. Arreguín-Sánchez (2012) documented changes in the Campeche Sound ecosystem associated with effects of climate, where the regime shift is clearly observed (Fig. 1.5). Likewise, Arreguín-Sánchez et al. (2015, 2019), del Monte-Luna et al. (2012), and Arreguín-Sánchez (2019) documented changes in diverse populations with different life histories in the same region (Fig. 1.6). Additionally, Sirot et al. (2017) reported metabolic differences in fish in the same region before,

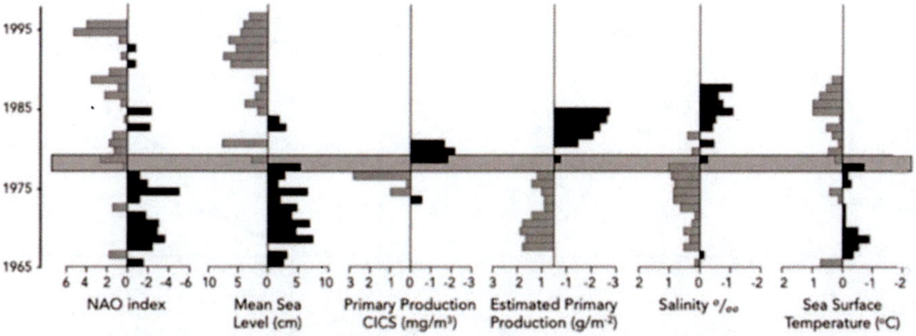

Fig. 1.5 Anomalies of different environmental variables and also for primary production showing the change of phase occurred at shift regime time around 1980s, for the Bank of Campeche in the Gulf of Mexico. NAO refers to the North Atlantic Oscillation. (Adapted from Arreguín-Sánchez 2012)

during, and after the regime shift, showing, for example, that the total protein content is lower towards the time of the regime shift (Fig. 1.7). The hypothesis formed from this information suggests global changes in the ecosystem that show relatively high productivity and that, at the time of the regime shift, represented by the end of a cooling period and the beginning of a warming period, global production decreased. This effect is noted in different trophic levels (Fig. 1.6) and is different for each; for example, the decreasing trend of primary production related to the decreasing abundance of short-living cycle species, such as the pink shrimp, *Farfantepenaeus duorarum* (Arreguín-Sánchez et al. 2015), or in long-lived species belonging to higher trophic levels, such as the grouper, *Epinephelus morio* (Arreguín-Sánchez et al. 2019). In contrast, an increase in the abundance of species such as the Mexican red octopus (*Octopus maya*) (Arreguín-Sánchez 2019) has also been documented; another example is a change in the response to climate, observed in the hawksbill turtle *Eretmochelys imbricata* (del Monte-Luna et al. 2012), a straddling species. The hypothesis essentially postulates that climate change affects the structure and function of the ecosystem; in this case, warming induced a decrease in global production over time, that is, gradually decreasing ecosystem's carrying capacity, where some populations adapted as a response, either increasing or decreasing in abundance. In particular, negatively affected species seem to strongly reduce their metabolic efficiency and their robustness towards their basal level, while adapting to the new condition, which will partially be restored after the environmental transition period, although not necessarily with the same metabolic efficiency and dynamics as in the previous period. If this hypothesis is documented with respect to individual metabolism, the processes described could illustrate the dynamics between biological hierarchies under conditions of environmental forcing. Based on Fig. 1.5, it is clear that management strategies should also be dynamically adapted to take advantage of, at each moment in time, the production capacity of species to ensure sustainability.

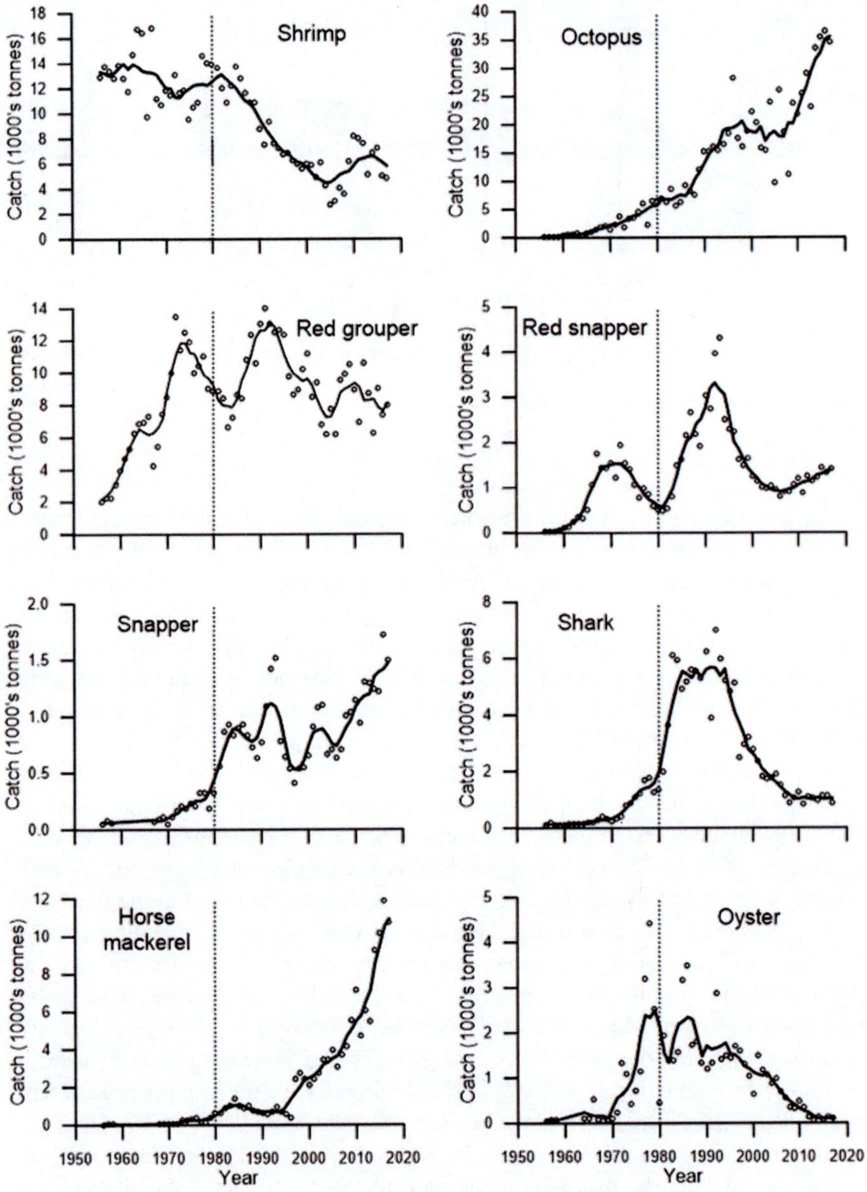

Fig. 1.6 Catch tendencies for several fish resources of the Campeche Bank in the Gulf of Mexico. Dashed vertical line represents the regime shift occurred approximately around 1980

According to Holling (1986), the development dynamics of each hierarchical level are shaped, from a holistic perspective, as an adaptive cycle of renewal (Holling 2001, see Chap. 2), in which, the magnitude of the change and its rate, within and between hierarchical levels, can be combined in various ways. From this

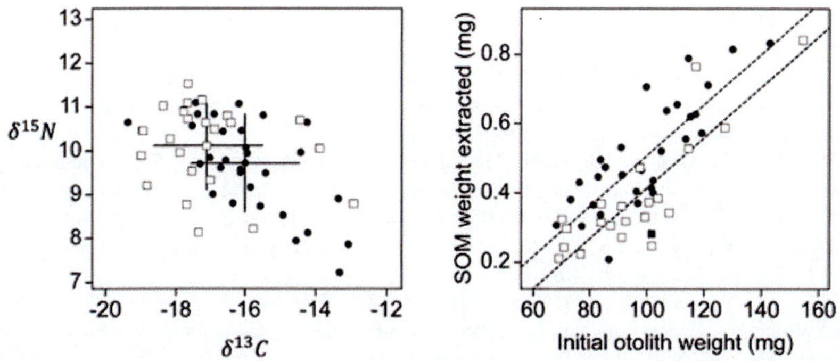

Fig. 1.7 Isotopic signatures (left) and soluble organic matter (SOM) in otoliths (right) of Bairdiella chrysoura in Laguna de Terminos, on the southern Gulf of Mexico; in two different periods, 1979–1980 (white squares), representing the end of the cooling period; and 2006–2007 (black circles), representing the present warming time. (Adapted from Sirot et al. 2017)

perspective, in response to environmental changes, organisms respond, populations adapt, and ecosystems are continuously reconfigured. In this way, the scientific approach to sustainable management should consider rules for controlling the use of resources based on attributes of the dynamics within and between biological hierarchies, whose relative weight in decision-making will be diverse.

References

Almpanidou, V., E. Katragkou, and A.D. Mazaris. 2018. The efficiency of phenological shifts as an adaptive response against climate change: a case study of loggerhead sea turtles (Caretta caretta) in the Mediterranean. *Mitigation and Adaptation Strategies for Global Change* 23: 1143–1158.

Arreguín-Sánchez, F. 2012. The dynamics linking biological hierarchies, fish stocks and ecosystems: implications for fisheries management. In *Models of the ecological hierarchy: from molecules to the ecosphere*, ed. F. Jordán and S.E. Jørgensen, 501–516. Philadelphia, PA: Elsevier B.V.

———. 2019. Climate change and the rise of the octopus fishery in the Campeche Bank, Mexico. *Regional Studies in Marine Science* 32: 100852.

Arreguín-Sánchez, F., E. Arcos, and E.A. Chávez. 2002. Flows of biomass and structure in an exploited benthic ecosystem in the gulf of California, Mexico. *Ecological Modeling* 156: 167–183.

Arreguín-Sánchez, F., P. Del Monte-Luna, and M.-J. Zetina-Rejón. 2015. Climate change effects on aquatic ecosystems and the challenge for fishery management: pink shrimp of the southern gulf of Mexico. *Fisheries* 40 (1): 15–19.

Arreguín-Sánchez, F., M.O. Albañez-Lucero, P. Del Monte-Luna, and M.-J. Zetina-Rejón. 2019. Fishery resource management challenges facing climate change. In *Mexican aquatic environments*, ed. A.L. Ibañez, 181–194. New York, NY: Springer.

Arreguín-Sánchez, F., P. Del Monte-Luna, M.O. Albañez-Lucero, M.-J. Zetina-Rejón, A. Tripp-Quezada, T.M. Ruiz-Barreiro, and J.C. Hernández-Padilla. 2021. How much biomass must remain in the sea after fishing to preserve ecosystem functioning? The case of the sardine fishery in the gulf of California, Mexico. In *Marine coastal ecosystems modelling and conservation*, ed. M. Ortiz and F. Jordán, 143–161pp. Cham: Springer Nature.

Barange, M., G. Merino, J.L. Blanchard, J. Scholtens, J. Harle, E.H. Allison, J.I. Allen, J. Holt, and S. Jennings. 2014. Impacts of climate change on marine ecosystem production in societies dependent on fisheries. *Nature Climate Change* 4: 211–216.

Bell, R.J., D.E. Richardson, J.A. Hare, P.D. Lynch, and P.S. Fratanton. 2015. Disentangling the effects of climate, abundance, and size on the distribution of marine fish: an example based on four stocks from the Northeast US shelf. *ICES Journal of Marine Science* 72 (5): 1311–1322.

Bell, J.D., R.A. Watson, and Y. Ye. 2017. Global fishing capacity and fishing effort from 1950 to 2012. *Fish and Fisheries* 18: 489–205.

Bell, J.D., A. Cisneros-Montemayor, Q. Hanich, J.E. Johnson, P. Lehodey, B.R. Moore, M.S. Pratchett, G. Reygondeau, I. Senina, J. Virdin, and C.C.C. Wabnitz. 2018. Adaptations to maintain the contributions of small-scale fisheries to food security in the Pacific Islands. *Marine Policy* 88: 303–314.

Birkmanis, C.A., J.J. Freer, L.W. Simmons, J.C. Partridge, and A.M.M. Sequeira. 2020. Future distribution of suitable habitat for pelagic sharks in Australia under climate change models. *Frontiers in Marine Science* 7: 570.

Brander, K.M. 2007. Global fish production and climate change. *Proceedings of the National Academy of Sciences of the United States of America* 104 (50): 19709–19714.

Brander, K. 2010. Impacts of climate change on fisheries. *Journal of Marine Systems* 79: 389–402.

Cabral, R.B., B.S. Halpern, S.E. Lester, C. White, S.D. Gaines, and C. Costello. 2019. Designing MpAs for food security in open-access fisheries. *Scientific Reports* 9: 8033.

Carlisle, A.B., R.E. Kochevar, M.C. Arostegui, J.E. Ganong, M. Castleton, J. Schratwieser, and B.A. Block. 2017. Influence of temperature and oxygen on the distribution of blue marlin (Makaira nigricans) in the Central Pacific. *Fisheries Oceanography* 26 (1): 34–48.

Chin, A., P.M. Kynew, T.I. Walkerz, and A.B. Mcaule. 2010. An integrated risk assessment for climate change: analysing the vulnerability of sharks and rays on Australia's Great Barrier Reef. *Global Change Biology* 16: 1936–1953.

Cochran, K. 2021. Reconciling sustainability, economic efficiency and equity in marine fisheries: has there been progress in the last 20 years? *Fish and Fisheries* 22: 298–323.

Coleman, F.C., and S.L. Williams. 2002. Overexploiting marine ecosystem engineers: potential consequences for biodiversity. *Trends in Ecology & Evolution* 17 (1): 40–44.

Coll, M., L.J. Shannon, K.M. Kleisner, M.J. Juan-Jordá, A. Bundy, A.G. Akoglu, D. Banaru, J.L. Boldt, M.F. Borges, A. Cook, I. Diallo, C. Fu, C. Fox, D. Gascuel, L.J. Gurney, T. Hattab, J.J. Heymans, D. Jouffre, B.R. Knight, S. Kucukavsar, S.I. Large, C. Lynam, A. Machias, K.N. Marshall, H. Masski, H. Ojaveer, C. Piroddi, J. Tam, D. Thiao, M. Thiaw, M.A. Torres, M. Travers-Trolet, K. Tsagarakis, I. Tuck, G.I. van der Meeren, D. Yemane, S.G. Zador, and Y.-J. Shin. 2016. Ecological indicators to capture the effects of fishing on biodiversity and conservation status of marine ecosystems. *Ecological Indicators* 60: 947–962.

Cury, P., and D. Pauly. 2020. Global marine fisheries: avoiding further collapses. In *Standing up for sustainable world*, ed. C. Henry, J. Rockström, and N. Stern. Cheltenham: Edward Elgar Publishing Limited.

Derville, S., L.G. Torres, R. Albertson, O. Andrews, S.C. Baker, P. Carzon, R. Constantine, M. Donoghue, C. Dutheil, A. Gannier, M. Oremus, M.M. Poole, J. Robbins, and C. Garrigue. 2018. Whales in warming water: assessing breeding habitat diversity and adaptability in Oceania's changing climate. *Global Change Biology* 25: 1466–1481.

Dufour, F., H. Arrizabalaga, X. Irigoien, and J. Santiago. 2010. Climate impacts on albacore and bluefin tunas migrations phenology and spatial distribution. *Progress in Oceanograpgy* 86: 283–290.

Eggleton, J.D., J. Depestele, A.J. Kenny, S.G. Bolam, and C. Garcia. 2018. How benthic habitats and bottom trawling affect trait composition in the T diet of seven demersal and benthivorous fish species in the North Sea. *Journal of Sea Research* 142: 132–146.

Engelhard, G.H., D.A. Righton, and J.K. Pinnegar. 2014. Climate change and fishing: a century of shifting distribution in North Sea cod. *Global Change Biology* 20: 2473–2483.

Erauskin-Extramiana, M., H. Arrizabalaga, A.J. Hobday, I. Arregui, H. Murua, and G. Chust. 2018. Large-scale distribution of tuna species in a warming ocean. *Global Change Biology* 25: 2043–2060.

Erauskin-Extramiana, M., M. Arrizabalaga, A. Cabré, R. Coelho, D. Rosa, L. Ibaibarriaga, and G. Chust. 2020. Are shifts in species distribution triggered by climate change? A swordfish case study. *Deep-Sea Research Part II* 175: 104666.

Fabinyi, M., W.H. Dressler, and M.D. Pido. 2017. Fish, trade and food security: moving beyond 'availability' discourse in marine conservation. *Human Ecology* 45: 177–188.

FAO. 2020. *El estado mundial de la pesca y la acuicultura 2020. La sostenibilidad en acción*. Roma: FAO. https://doi.org/10.4060/ca9229es.

Fariñas-Franco, J.M., A.L. Allcock, and D. Roberts. 2018. Protection alone may not promote natural recovery of biogenic habitats of high biodiversity damaged by mobile fishing gears. *Marine Environmental Research* 135: 18–28.

Gilman, E., A. Perez Roda, T. Huntington, S.J. Kennelly, P. Suuronen, M. Chaloupka, and P.A.H. Medley. 2020. Benchmarking global fisheries discards. *Scientific Reports Nature Research* 10: 14017.

Glantz, M.H., ed. 1992. *Climate variability, climate change, and fisheries*. Cambridge: Cambridge University Press.

Hazen, E.L., S. Jorgensen, R.R. Rykaczewski, S.J. Bograd, D.G. Foley, I.D. Jonsen, S.A. Shaffer, J.P. Dunne, D.P. Costa, L.B. Crowder, and B.A. Block. 2012. Predicted habitat shifts of Pacific top predators in a changing climate. *Nature Climate Change* 3: 234–238.

Holland, D.S. 2010. Markets, pooling and insurance for managing bycatch in fisheries. *Ecological Economics* 70: 121–133.

Holling, C.S. 1986. Adaptive environmental management. *Environment: Science and Policy for Sustainable Development* 28: 9–39.

———. 2001. Understanding the complexity of economic, ecological, and social systems. *Ecosystems* 4: 390–405.

Holling, C.S., and L.H. Gunderson. 2002. Resilience and adaptive cycles. In *Panarchy: understanding transformations in human and natural systems*, ed. L. Gunderson and C.S. Holling, 25–62. Washington, DC: Island Press.

Kennedy, J. J., N.A. Rayner, R.O. Smith, D.E. Parker, and M. Saunby. 2011a. Reassessing biases and other uncertainties in sea surface temperature observations measured in situ since 1850: 1. Measurement and sampling uncertainties. Journal of Geophysical Research 116(D14) 10.1029/2010JD015218

Kennedy, J. J., N.A. Rayner, R.O. Smith, D.E. Parker, and M. Saunby. 2011b. Reassessing biases and other uncertainties in sea surface temperature observations measured in situ since 1850: 2. Biases and homogenization. Journal of Geophysical Research 116(D14) 10.1029/2010JD015220

Kennelly, S.J., and M.K. Broadhurst. 2021. A review of bycatch reduction in demersal fish trawl. *Fish and Fisheries* 31: 289–318.

Kopmoroske, L.M., and R.L. Lewison. 2015. Addressing fisheries bycatch in a changing world. *Frontiers in Marine Science* 2 (83): 1–11.

Lambert, G.I., S. Jennings, M.J. Kaiser, T.W. Davies, and J.G. Hiddink. 2014. Quantifying recovery rates and resilience of seabed habitats impacted by bottom fishing. *Journal of Applied Ecology* 51: 1326–1336.

Lehodey, P., J. Alheit, M. Barange, T. Baumgartner, G. Beaugrand, K. Drinkwater, J.M. Fromentin, S.R. Hare, G. Ottersen, R.I. Perry, C. Roy, C.D. Van Der Lingen, and F. Werner. 2006. Climate variability, fish, and fisheries. *Journal of Climate* 19: 5009–5030.

Link, J.S., and R.A. Watson. 2019. Global ecosystem overfishing: clear delineation within real limits to production. *Science Advances* 5: eaa0474.

Link, J.S., J.A. Nye, and J.A. Hare. 2011. Guidelines for incorporating fish distribution shifts into a fisheries management context. *Fish and Fisheries* 12: 461–469.

Link, J.S., R.A. Watson, F. Pranovic, and S. Libralato. 2020. Comparative production of fisheries yields and ecosystem T overfishing in African Large Marine Ecosystems. *Environmental Development* 36: 100529.

Lluch-Belda, D., S. Lluch-Cota, D. Lluch-Cota, and S. Hernández-Vázquez. 1999. La variabilidad oceánica interanual y su impacto en las pesquerías. *Revista de la Sociedad Mexicana de Historia Natural* 49: 219–227.

Lluch-Cota, S.E., A. Parés-Sierra, V.O. Magaña-Rueda, F. Arreguín-Sánchez, G. Bazzino, H. Herrera-Cervantes, and D. Lluch-Belda. 2010. Changing climate in the Gulf of California. *Progress in Oceanography* 87: 114–136.

Lluch-Cota, S.E., F. Arreguín-Sánchez, C.J. Salvadeo, and P. Del Monte Luna. 2018. Climate change impacts, vulnerabilities and adaptations: north-east tropical Pacific marine fisheries. In *Impacts of climate change on fisheries and aquaculture: synthesis of current knowledge, adaptation and mitigation options*, FAO Fisheries and Aquaculture technical paper no. 627, ed. M. Barange, T. Bahri, M.C.M. Beveridge, K.L. Cochrane, S. Funge-Smith, and F. Poulain, 207–218. Rome: FAO. 628 p.

Loring, P.A., D.V. Fazzino II, M. Agapito, R. Chuenpagdee, G. Gannon, and M. Isaacs. 2019. Fish and food security in small-scale fisheries. In *Transdisciplinarity for small-scale fisheries governance*, MARE Publication series 21, ed. R. Chuenpagdee and S. Jentoft, 55–73. Cham: Springer Nature.

Lotze, H.K., M. Coll, and J.A. Dunne. 2011. Historical changes in marine resources, food-web structure and ecosystem functioning in the Adriatic Sea, Mediterranean. *Ecosystems* 14: 198–222.

Martínez-Arroyo, A., S. Manzanilla-Naim, and J. Zavala-Hidalgo. 2011. Vulnerability to climate change of marine and coastal fisheries in México. *Atmosfera* 24 (1): 103–123.

Martins-Vieira, M.L., C.L. Alexandre-de-Lima, J.R. Botelho-de-Souza, and J.L. Leão-Feitosa. 2020. Effects of beach seine fishing on the biodiversity of seagrass fish assemblages. *Regional Studies in Marine Science* 40: 101527.

Meynecke, J.O., E. Seyboth, J. De Bie, J.L. Menzel-Barraqueta, A. Chama, S.P. Dey, S.B. Lee, V. Tulloch, M. Vichi, K. Findlay, A.N. Roychoudhury, and B. Mackey. 2020. Responses of humpback whales to a changing climate in the Southern Hemisphere: priorities for research efforts. *Marine Ecology* 41: e12616.

del Monte-Luna, P., V. Guzmán-Hernández, E.A. Cuevas, F. Arreguín-Sánchez, and D. Lluch-Belda. 2012. Effect of North Atlantic climate variability on hawksbill turtles in the Southern Gulf of Mexico. *Journal of Experimental Marine Biology and Ecology* 412: 103–109.

Patino-Martinez, J., A. Marco, L. Quiñones, and L. Hawkes. 2011. A potential tool to mitigate the impacts of climate change to the Caribbean leatherback sea turtle. *Global Change Biology* 18: 401–411.

Pauly, D., J. Alder, E. Bennett, V. Christensen, P. Tyedmers, and R. Watson. 2003. The future of fisheries. *Science* 302: 1359–1361.

Pepin, P., J. King, C. Holt, H. Gurney-Smith, N. Shackell, K. Hedges, and A. Bundy. 2020. Incorporating climate, oceanographic and ecological change considerations into population assessments: a review of Fisheries and Oceans Canada's science advisory process. In *DFO the Canadian Science Advisory Secretariat research document*. 2019/043. iv + 66 p.

Poloczanska, E., C.J. Limpus, and G.C. Hays. 2009. Vulnerability of marine turtles to climate change. *Advances in Marine Biology* 56: 151–211.

Porobic, J., E.A. Fulton, C. Parada, S. Frusher, B. Ernst, and P. Manríquez. 2019. The impact of fishing on a highly vulnerable ecosystem, the case of Juan Fernández Ridge ecosystem. *PLoS One* 14 (2): e0212485.

Ragheb, E., E.H.K. Akel, and S.I. Rizkalla. 2019. Analyses of the non-target catch from the Egyptian Mediterranean trawlers, off Port Said. *Egyptian Journal of Aquatic Research* 45: 239–246.

Rijnsdorp, A.D., S.G. Bolam, C. Garcia, J.G. Hiddink, N.T. Hintzen, P.D. Van Denderen, and T. Van Kooten. 2018. Estimating sensitivity of seabed habitats to disturbance by bottom trawling based on the longevity of benthic fauna. *Ecological Applications* 28 (5): 1302–1312.

Salcido-Guevara, L.A., P. del Monte-Luna, F. Arreguín-Sánchez, and V.H. Cruz-Escalona. 2012. Potential ecosystem level effects of a shrimp trawling fishery in La Paz Bay, Mexico. *Open Journal of Marine Science* 2: 85–89.

Schaefer, M.B. 1954. Some aspects of the dynamics of populations important to the management of commercial marine fisheries. *Bulletin - Inter-American Tropical Tuna Commission* 1: 27–56.

———. 1957. A study of the dynamics of the fishery for yellowfin tuna in the eastern tropical Pacific Ocean. *Bulletin - Inter-American Tropical Tuna Commission* 2: 247–285.

Scheinin, A.P., D. Kerem, C.D. Macleod, M. Gazo, C.A. Chicote, and M. Castellote. 2011. Gray whale (Eschrichtius robustus) in the Mediterranean Sea: anomalous event or early sign of climate-driven distribution change? *Marine Biodiversity Records* 4: e28.

Sirot, C., P. Grønkjær, J.B. Pedersen, J. Panfili, M. Zetina-Rejon, A. Tripp-Valdez, J. Ramos-Miranda, D. Flores-Hernandez, A. Sosa-Lopez, and A.M. Darnaude. 2017. Using otolith organic matter to detect diet shifts in Bardiella chrysoura, during a period of environmental changes. *Marine Ecology Progress Series* 575: 137–152.

Srinivasan, U.T., R. Watson, and U.R. Sumaila. 2012. Global fisheries losses at the exclusive economic zone level, 1950 to present. *Marine Policy* 36: 544–549.

Stacey, N., D.J. Steenbergen, J. Clifton, and G. Acciaioli. 2018. Understanding social wellbeing and values of small-scale fisheries amongst the Sama-Bajau of archipelagic southeast Asia. In *Social wellbeing and the values of small-scale fisheries*, ed. D.S. Johnson, T.G. Acott, N. Stacey, and J. Urquhart, 97–123pp. Heidelberg: Springer.

Suuronen, P., and E. Gilman. 2020. Monitoring and managing fisheries discards: new technologies and approaches. *Marine Policy* 116: 103554.

Takeshige, A., M. Miyamoto, Y. Narimatsu, S. Yonezaki, and M. Kiyota. 2021. Evaluation of impacts of bottom fishing on demersal habitats: a case study off the Pacific coast of north-eastern Japan. *Fisheries Research* 238: 105916.

Teh, L.C.L., and D. Pauly. 2018. Who brings in the fish? The relative contribution of small-scale and industrial fisheries to food security in Southeast Asia. *Frontiers in Marine Science* 5: 44.

Thrush, S.F., J.E. Hewitt, G.A. Funnell, V.J. Cummings, J. Ellis, D. Schultz, D. Talley, and A. Norkko. 2001. Fishing disturbance and marine biodiversity: role of habitat structure in simple soft-sediment systems. *Marine Ecology Progress Series* 221: 255–264.

United Nations General Assembly. 1987. *Report of the World Commission on Environment and Development: our common future*. Oslo: United Nations General Assembly, Development and International Co-operation: Environment.

Vargas-López, V.G., F. Vergara-Solana, and F. Arreguín-Sánchez. 2021. Effect of environmental variability on the individual growth of yellow abalone (Haliotis corrugata) and blue abalone (Haliotis fulgens) in the Mexican Pacific. *Regional Studies in Marine Science* 46: 101877.

Woodworth-Jefcoats, P.A., J.J. Polovina, and A.C. Drazen. 2017. Climate change is projected to reduce carrying capacity and redistribute species richness in North Pacific pelagic marine ecosystem. *Global Change Biology* 23: 1000–1008.

Worm, B., R. Hilborn, R. Baum, J.K. Trevor, A. Branch, J.S. Collie, C. Costello, M.J. Fogarty, E.A. Fulton, J.A. Hutchings, S. Jennings, O.P. Jensen, H.K. Lotze, P.M. Mace, T.R. McClanahan, C. Minto, S.R. Palumbi, A.M. Parma, D. Ricard, A.A. Rosenberg, R. Watson, and D. Zeller. 2009. Rebuilding global fisheries. *Science* 325: 578–585.

Ye, Y., and F. Carocci. 2018. Control mechanisms and ecosystem-based management of fishery production. *Fish and Fisheries* 20: 15–24.

Ye, Y., K. Cochrane, G. Bianchi, R. Willmann, J. Majkowski, M. Tandstad, and F. Carocci. 2013. Rebuilding global fisheries: the World Summit Goal, costs and benefits. *Fish and Fisheries* 14: 174–185.

Zhou, S., A.D.M. Smith, and M. Fuller. 2011. Quantitative ecological risk assessment for fishing effects on diverse data-poor non-target species in a multi-sector and multi-gear fishery. *Fisheries Research* 112: 168–178.

Chapter 2
Conventional Fisheries Management and the Need for an Ecosystem Approach

The Need of Knowledge for Management

There is a general concern regarding the adequate use of living marine resources, and for this, scientific knowledge has been adapted and applied according to the needs of the industry, either to generate foreign exchange or economic income or to generate employment or food. Thus, for the first decades of the twentieth century, simply determining the available biomass and the location of exploitable stocks was sufficient; with this, access conditions, quotas, closures, harvest rates, etc. could be defined. Subsequently, as the fishing sector develops, capitalizes, and grows with the increase in fishing capacity and fishing power, it is recognized not only that resources are limited but also that fishing capacity can exceed these limits and lead to the overfishing and collapse of the fish resources. Faced with these needs, greater and better knowledge of fisheries is required. Fisheries science is developed based upon the population dynamics, proposing mathematical models that, while necessarily are a simplification of nature, strongly support decision-making in the use of living marine resources. Initially, it was enough to determine the available biomass and production capacity, but eventually, with a greater fishing pressure, it became necessary to know the specific impact of fishing on the population structure; for example, to ensure that enough mature females survive to reproduce and recover the losses due to fishing. This kind of knowledge became of relevance towards the end of the twentieth century and the beginning of the twenty-first century, being essential for many fisheries that required precise estimates incorporating a statistical component that allowed to deal with variability and uncertainty into the models to formally support risk in the decision-making process. In this context, Smith (1994) reviews changes in fishing during these periods, as well as the evolution of fisheries science geared towards industry during the second half of the twentieth century.

In the current context, fishing intensity and technological development provide a highly efficient fishing power, being causes for concern due to the deterioration reported for various fisheries around the world (Clark and Munro 2002; Zeller and

F. Arreguín-Sánchez, *Holistic Approach to Ecosystem-Based Fisheries Management*, https://doi.org/10.1007/978-3-030-96847-2_2

Pauly 2019; FAO 2020); and although this concern is justifiable, it is also known that the state of resources is also due to sources of variation other than fishing and that, in turn, affect the production capacity of the fish stocks; such is the case of the effects of the changes in climate patterns, rivers' discharges, upwellings, winds, hurricanes, among other natural factors; which could include the consequences of the greenhouse effects, ocean acidification and other human activities that affect marine regions, such as deforestation, and dumping of toxic substances and materials into the sea, among other activities. Currently, given these circumstances, scientific knowledge applied to resource management should incorporate the main sources of variation of the abundance of exploited resources, including fishing, the environmental variables, as well as the dynamic within and between the different hierarchical levels of biological organization (Arreguín-Sánchez 2012).

Regarding climate change, as a process that induces changes in trends in natural systems, consider for a moment the sea surface temperature anomalies in the last century and a half as an acceptable indicator of climate change that affects marine ecosystems to analyze the historical trend of the development of fisheries science (Fig. 2.1). Population-based fishery models were conceptually developed, and strongly evolved, since the 1920s, based on Verhulst (1838) equation, and the work of Baranov (1918). A few decades later, surplus production models were proposed (Schaefer 1954; Pella and Tomlinson 1969; Fox 1970), the yield-per-recruit model (Beverton and Holt 1957), and size- and age-structured stock assessment models (Gulland 1965; Pope 1972; Jones 1981; Pope and Shepherd 1982, 1985; Pauly and Morgan 1987). All these models explicitly or implicitly assume a stable carrying capacity of populations and consequently of ecosystems. The main consequence of this assumption is that fishing is the main source of variation influencing abundance in exploited populations, an assumption that could have been acceptable during the 1920–1980 period; but not after that, as suggested by the trend in temperature anomalies shown in Fig. 2.1.

However, fishery science faced, at this point, two particular situations: the fact that model predictions fell short of the expected outcomes, with not always desired consequences for the management of fish resources; and that the global catch trend began to stabilize or decline. These conditions, supported by the available science at the time, triggered four approaches: (1) the management of fully exploited fisheries, that is, of fish resources exploited at their maximum production capacity, (2) the recovery of overexploited fish resources, and (3) the explicit recognition of the effects of the environment on the exploited fish stocks, and (4) the effect of fishing on non-target species and habitats, giving rise to the so-called ecosystem approach that includes both climate effects and the effect of biological interdependencies, and in particular, but not exclusively, trophic relationships.

For the first two approaches, it is expected that science provides more accurate information in terms of fishing mortality and available biomass. Thus, towards the 1990s, the incorporation of statistical tools in population models took off, in particular, approaches based on maximum likelihood estimates, Bayes theorem, and uncertainty and risk analysis including probabilities of occurrence and nonoccurrence of events or processes (Punt and Hilborn 1997, 2001; Lele et al.

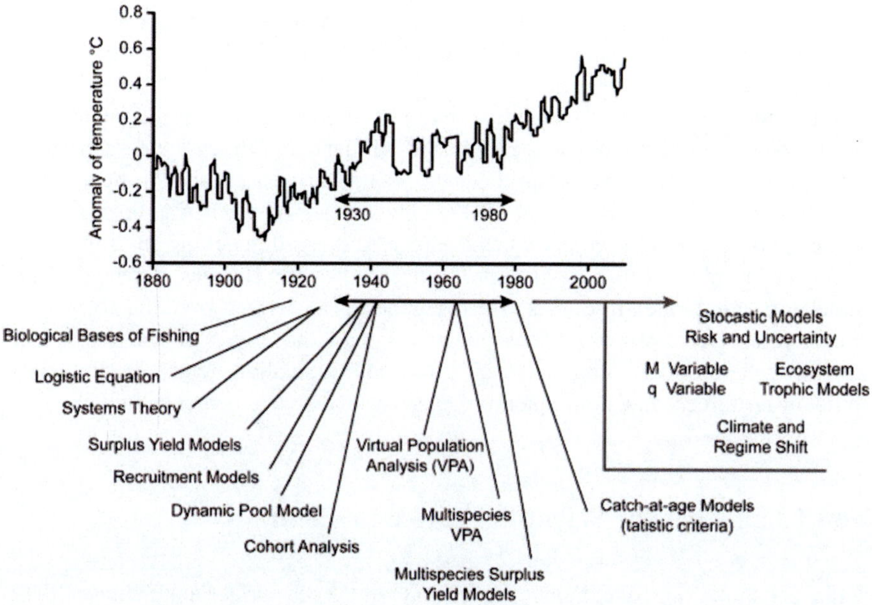

Fig. 2.1 Time series of the anomaly of the global sea surface temperature and of the significative scientific contributions to fisheries along this time. Note the stable period from 1930 until 1980, when the regime change occurred, and a warming period initiate. Most stock assessment theoretical basis and methods were developed during this period, and, most of them, under the explicitly or implicitly, assumed such stability, as consequence, that fishing was the main driver of changes in abundance. Since fishing pressure became more intensive, precision of methods also became as necessary; this meant more details in about the ages/sizes structure and, statistically, a better description of estimations to diminish risks in decision-making processes. After the regime shift in about 1980, environment, represented by anomaly in temperature, becomes a significant driver of abundance, in addition to fishing and, in some cases, even with a stronger effect. After this time, scientific research increased to provide more precise information, and particularly incorporating climate effects in management. Even when such knowledge was impulsed by the need to solve immediate problems, globally they consider that in natural systems individuals suffer changes in their metabolism affecting populations and that such populations adapt through their dynamics to such changes and, as consequence, ecosystem reconfigure themselves; all this as a continuous process that need to be scientifically described and assessed to give support, with solid knowledge, to an adaptability-based management policy

2007; Hilborn 2012; Stewart et al. 2013; Doll and Jacquemin 2019), as well as the independent estimation of patterns of change in parameters such as catchability (Arreguín-Sánchez 1996; Arreguín-Sánchez and Pitcher 1999) and natural mortality (Gulland 1987a; Caddy 1996; Martínez-Aguilar et al. 2005. Torrejón-Magallanes et al. 2021). These methods and procedures aim to achieve a more accurate estimate of the state of the resources, fishing mortality, harvest rates, exploitable biomass, and consequently, allowable catch (Anderson 2010; Shephard et al. 2015; Richter et al. 2018). However, despite these advances in science applied to fish resource management, the key assumption continued to be that fishing was the main driver of changes in abundance and therefore, that with an adequate control of fishing capacity, the expected response of resource recovery would be achieved. However, this has not been the case in recent decades.

While the importance of variability due to natural causes and of the interdependencies between populations was recognized (Gulland 1987b, c; Hannesson 1983; Murawski 1991, 1993; Arreguín-Sánchez 2000; Christensen and Walters 2005; Morishita 2008; Pérez-Pérez et al. 2017; Lewis et al. 2019), it was not until the beginning of the present century when awareness emerged regarding environmental forcing and how it affects biological production and causes changes in trends for the abundance of exploited resources. This awareness was derived from the accumulation of scientific observations; that is, it was until the effects of environmental variables where documented, and their effects on population trends, on the availability of resources, and, consequently, on ecosystems, were demonstrated.

New Challenges for Scientific Advice to Fishing

Figure 2.1 shows a gradual monotonic increase in the sea surface temperature (SST) anomalies from the beginning of the 1980s, and again, if it is assumed that this trend is an indicator of the climate change that affect marine ecosystems, it is then easy to understand that natural systems respond to these changes with different arrangements over time and space as a result of a process of continuous adaptability, a process in which the ecosystem evolves and is reconfigured under an environmental forcing regime that also induces continuous and gradual changes in its functioning and probably on structural processes. Given this circumstance, it is clear that the science developed and used successfully in the decades prior to 1980 cannot be expected to be a sufficient scientific tool to support current management decisions with expected success. This implies a radical change in the paradigms of science that need to be changed to address these new circumstances. In essence, in terms of the mathematical models available until that time, the global assumptions that constitute the basis of these models ceased to be reasonably valid as a frame of reference. This situation clearly defines the current need for a change in focus with respect to conventional science. In this context, it is clear that environmental forcing and the characteristics defined by continuous changes in ecosystem function and organization are sources of significant variation in the biomass availability of particular fishery resources. What is expected is that the incorporation of knowledge of these processes contributes to explaining the observed variability, in addition to fishing, on the abundance of resources. In this way, fishing, as the only source of variation that is under our control, could be directed towards a particular objective with a high probability of success. In this way, explaining variability according to the processes of environmental forcing and changes in ecosystem function and organization will allow us to separate the real effects of fishing and, in turn, increase the probability of successful fishing controls by applying strategies that are adaptable to changes of the ecosystem conditions.

The scientific transition from one state to another, that is, from a state of relative stability to one of continuous trend changes, has not been easy. Science requires evidence, for which time has to pass to get it, demonstrating the changes and effects of climate change and the continuous evolution of ecosystems and populations, and the aspects that has not been documented for many fisheries. In this transition, several fisheries were overfished, most likely involuntarily. Imagine a population exploited to the maximum of its biological production capacity, where each year the amount of biomass loss due to fishing is replenished through surplus production, maintaining sustained yields through the fishing capacity of the fleet. Now, imagine environmental forcing that causes a decrease in production capacity; as a consequence, the losses are not completely replenished, and this cumulative effect over time causes the abundance of the resource to decrease such that the resource is increasingly unable to replenish that loss differential. As populations naturally show some interannual variability, the short-term changes in yields, and even in population abundance, are imperceptible to the industry until these declines occur over a period of several years. If a population is exploited to its maximum productive capacity and the stock decreases continuously over time as a result of environmental forcing, the resource will be overexploited, resulting, in extreme cases, in the collapse of some fisheries (i.e., Arreguín-Sánchez et al. 2015). This condition occurs when the fishing capacity does not adapt over time to the new stock size and consequently to the new production capacity. This process led to two scientific recommendations; one involves trying to adapt fishing effort to estimated stocks sizes using population models, with the drawback that this approach maintains the same assumptions of the origin models, that is, a stable carrying capacity. The second recommendation involves changing the scientific paradigm to advise the fishing industry through an ecosystem-based approach, which is nothing more than the explicit recognition of the effect of the environment and the dynamics within and between biological hierarchies (populations and ecosystem), in the context of the continuous changes that natural systems are experiencing.

The problem at present is that in various cases, scientific advice continues to rest on the assumption of a stable carrying capacity of populations and ecosystems. This approach is already insufficient for many fisheries, and in this context, it is necessary for scientific endeavors to evaluate and separate the sources of variation in the abundance of resources, especially those that define trends over time. The nonexplicit consideration of the significant sources of variation that affect the production capacity and the stock abundance involves a higher uncertainty regarding the expected effect of management measures, and also a higher risk to fail; by not doing so, the variation due to causes other than fishing will be attributed to, within the modeling processes, effects of fishing. When this occurs in the management, measures do not result in the expected effects, uncertainty is generated in the productive sector, and worse, when this persists over time, distrust of institutions and eventually illegal activity and ungovernability are generated.

An aspect of interest to consider is that the continuous reconfiguration of ecosystems, in response to environmental forcing, induces changes in the thermodynamic balance of ecosystems; that is, changes in the abundance of populations affect

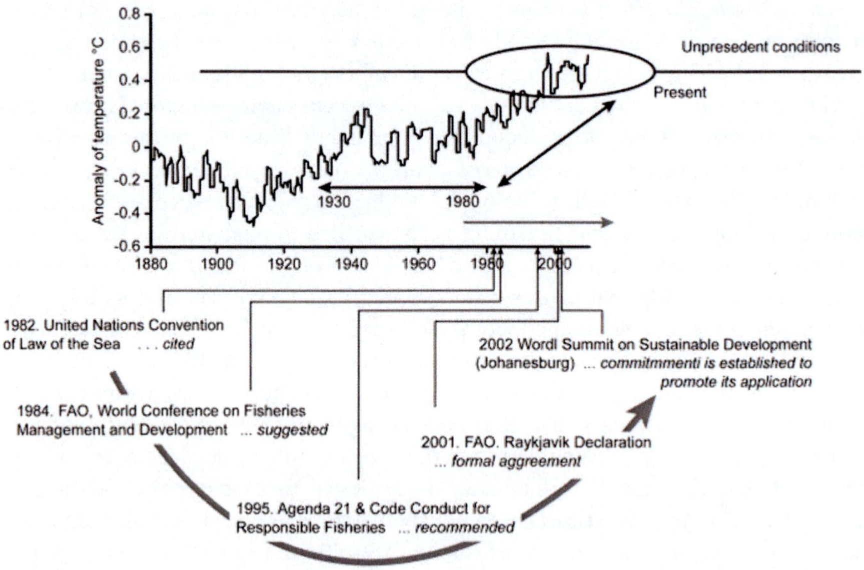

Fig. 2.2 Time series describing the evolution of the role of the intergovernmental entities to adopt formally climate change as a phenomenon affecting fisheries among other natural resources; and the need to attend it integrating the impact in the management policies within the countries. Note that application of policies in the countries was formally adopted as commitment until 2002; 20 years after climate regime shift. Such lag is mostly explained because of the need of hard evidence on impacts and effects which needed, in turn, time to be documented. Note that the present configuration of the ecosystems probably present processes and dynamics under unprecedented conditions

the predator–prey relationships that define the energy flows and, subsequently, the structure, organization, and function of an ecosystem. This thermodynamic balance is of great relevance because, at the ecosystem level, it defines attributes such as the resilience and self-organizing capacity of the system, among others. Regarding the scientific community putting the ecosystem approach into practice, including for climate change, the process has been slower than desirable. Consider Fig. 2.2, and again suppose that trend for temperature anomalies is an acceptable proxy of the state of the environment. From the perspective of the world's governments in 1982, the United Nations Convention on the Law of the Sea (Levy 1993) mentions the relevance that climate change could have on nature and society, claiming that its primary origin is identified by the regime shift (Young 1982; Cicin-Sain and Knecht 1989), where a cooling period ended and a warming period began. In 1984 at the World Conference on Fisheries Management and Development (Bojó 1985), it was suggested to address the process of climate change; however, and probably because there was still not enough evidence of the potential effect of this phenomenon, a decade passed, and it was not until 1995 that an explicit recommendation for countries appeared in Agenda 21 in 1992 (UN 1992; Sitarz 1993) and in the Code of Conduct for Responsible Fisheries (FAO 1995) to formally consider this phenomenon. Later, in 2001, in the Reykjavik Declaration (2001), the participating countries agreed on the need to address aspects related to climate change, and in 2002, at the World Summit on Sustainable Development in Johannesburg (UN 2002), in an explicit recognition of the relevance of the process, the countries committed to implementing policies that consider the causes of climate change and how to deal with. In the sequence of events shown in Fig. 2.2, it can be seen that it took several years for science to formally provide evidence of the effects of climate change, after which countries became aware of the reality of the phenomenon and its consequences on nature and society.

Currently, resources and ecosystems face unprecedented conditions in the history of humanity because of the combination of environmental processes, such as global climate change, changes in local environmental regimes, acidification processes and changes in ocean circulation patterns, and unprecedented high-intensity fishing, which includes technological advances in fishing power. In the first case, scientific evidence suggests that ecosystems and, consequently, populations have been influenced by a continuous warming process for almost four decades, that is, by a regime shift (Dinga et al. 2017; Gaines et al. 2018; Phillips and Perez-Ramírez 2018; Lam et al. 2020; Mendenhall et al. 2020). Ecosystems, in this process of continuous change, are also continuously rearranged, and within them, the populations change. The warming process corresponds to the rising phase of a cyclical process in which, in a typical period, maximum values would occur towards the beginning of the 2020s; however, this condition could last longer when coupled with the effect of warming caused by human activity, such as the accumulation of greenhouse gases, a situation that, at this time, is not predictable. The second aspect refers to fishing intensity. According to the FAO (2020), more than 60% of the living marine resources are fully exploited, while approximately 20% are in a state of overfishing, for which, in principle, recovery strategies have been defined; 16% have collapsed, while the remaining 4% corresponds to developing fisheries. These unprecedented conditions also require an unprecedented scientific paradigm shift that, with the necessary knowledge about populations and ecosystems contends with this dynamic and allows a process of adaptability in management that promotes, within the dynamics of natural systems, sustainability.

Of course, in the climate context, there are different recognized sources of variation that affect fishery resources, such as PDO, MEI, NAO, AMO, and ENSO, each with their causes and frequencies acting fundamentally at the regional level, but being relevant at the local level. The complexity of these processes can complicate the scientific predictions of stocks abundance because the combination of effects and frequencies generates particular conditions that, in turn, affect resources and ecosystems in a specific way, generating differential responses that mask or combine each other and are therefore difficult to predict. A clear example of the complexity of this type of process is the Gulf of California (Lluch-Cota et al. 2010, 2018), where, for instance, the influence of El Niño effects determine the influx of warm water masses that negatively affect primary phytoplankton production. However, under the influence of El Niño phenomena, easterly winds also intensify, which in turn causes upwelling in the continental coast of the gulf, promoting primary productivity. The net effect on biological production depends on the balance of the intensity of these processes; however, in addition, other types of significant environmental effects have been reported, such as PDO and PNA (Lluch-Cota et al. 2018), that, given the specific processes and the scales of their frequencies, affect, in a combined way, the biology of populations in various ways and whose responses vary according to their life histories. In essence, what this type of evidence suggests is that natural systems do not uniformly respond in similar patterns to these environmental stimuli but also depend, to a large extent, on local or regional processes.

Additionally, with climate change, warmer sea temperatures are accompanied by other events, such as reduced oxygen levels and ocean acidification caused by an increase in the CO_2 concentration, phenomena that represent limits to physiological mechanisms of individuals and therefore of populations and ecosystems. The response of each ecological hierarchy, according to the biological and ecological conditions imposed by their life histories and their ability to adapt, affects biodiversity, species distribution, and the efficiency of key processes such as reproduction and biological interdependencies, independently of fishing. Given the possible combinations of biological and ecological processes underlying the responses to climate effects, it is easy to understand why conventional management measures, typically passive, do not work as expected lacking environmental stability. The understanding of these processes makes clear the need to change the approach of scientific recommendations to an ecosystem perspective with adaptive dynamic management strategies.

Ecosystem Approach to Fisheries: Conventional Approach

In general, resource management objectives are defined for economic and social reasons and typically based on the status of the exploited fish resource. Institutional organization, as part of the governance system, is designed on this basis; that is, resources are generally managed independently of each other despite belonging to the same ecosystem; this implicitly assumes the absence of biological and ecological interdependencies. For example, for the management of a coastal pelagic species, management actions directed to a demersal species are not considered, even when both belong to the same ecosystem, or even within the same coastal area in the continental shelf. The conventional concept of ecosystem approach to fisheries management, EAFM, has been applied basically, by considering the direct effects of fishing on the ecosystem but not the indirect effects. Figure 2.3 illustrates this scenario, where each fishery focused on a different target resource considers only the impacts on those components of a portion of the ecosystem to which it accesses or with which the fishery is directly related. In this way, management actions are geared "from the fish stock towards the ecosystem" and are generally focused to the direct effects of fishing or on the direct effects on other fisheries (Jennings and Kaiser 1998; Tudela 2004; Ortuño-Crespo and Dunn 2017; Liang and Pauly 2017, 2020; Xing et al. 2021); as well as the reduction in bycatch, the biodiversity conservation or threatened species; or the impact on habitat, or pollution associated with fishing

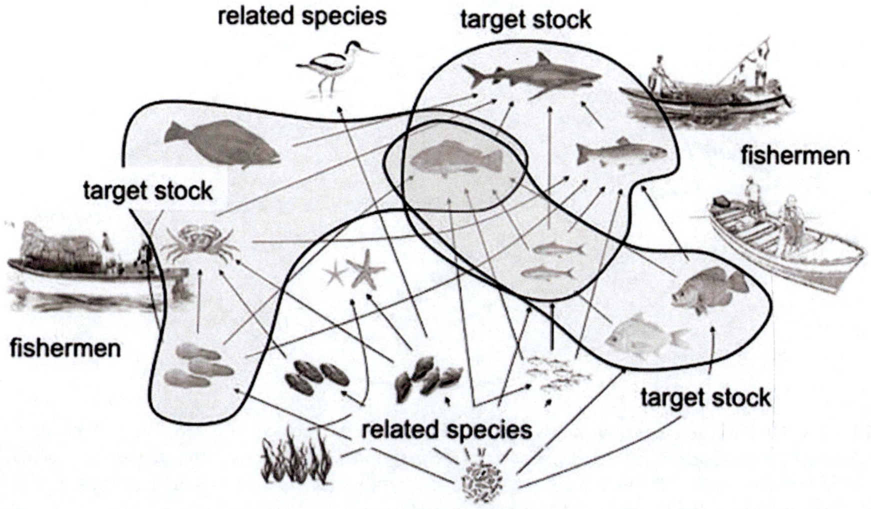

Fig. 2.3 Diagram representing the prevailing current approach to ecosystem-based fisheries management, "from individual stocks to ecosystem" approach. In this case the impact of fishing can be directly estimated for target speciesand bycatch; but this approach does not consider changes in the function and organization of ecosystems. The diagram exemplifies the fishing of clams, intermediate trophic level fish or sharks, representative of high trophic levels. Its impacts have a given scope within the ecosystem, but the ecosystem is never considered as a functional unit of nature

operations; and, recently, on species considered as forage in the ecosystem and that are exploited as target fish resources (Schmiing et al. 2015; Coll et al. 2016; Richardson et al. 2017; Fariñas-Franco et al. 2018; Arreguín-Sánchez et al. 2021).

The current application of the EAFM does not provide information on aspects of the ecosystem as a natural unit; that is, fishing control criteria are not generated from indicators of holistic attributes of ecosystems. These indicators should be directed to expressing attributes related to maintaining the organization and function of the ecosystem, its resilience, productive capacity, and structural cohesion, among others; all of them being attributes that together and balanced gives the ecosystem its ability to endure over time and to sustainably maintain its global production capacity and therefore of its use. Currently, ecosystems go through a process of continuous reconfiguration, promoted by environmental change patterns, in such a way that the conservation of holistic attributes and the possibility of expressing them as reference indicators represent one of the perspectives of greatest scientific interest; all of this, of course, under a dynamic sustainability criterion and, therefore, independent of the intrinsic characteristics of the individual components of the ecosystem. In this sense, holistic attributes can be expressed as performance indicators for applying control measures to fisheries for the use of explicit target resources, thus generating a "from the ecosystem to the fish stocks" management perspective (Fig. 2.4).

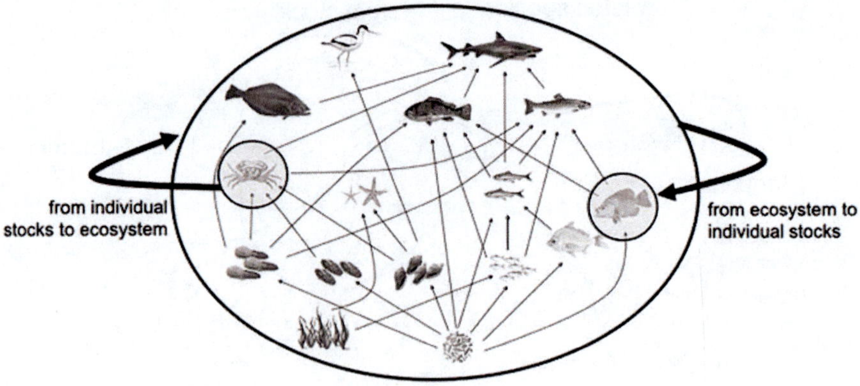

Fig. 2.4 Diagram representing the two global tendencies for ecosystem-based fisheries management; the present conventional approach "from the individual populations to the ecosystem" versus "from the ecosystem to the individual populations." The former was explained in detail in Fig. 2.3; the last consists in the definition of individual harvest rates derived from the ecosystem attributes related with their maintenance of their organization and function

References

Anderson, L.G. 2010. Setting allowable catch levels within a stock rebuilding plan. In *The economics of rebuilding fisheries: workshop proceedings*. Paris: OECD Publishing.

Arreguiín-Sánchez, F. 2000. Octopus – red grouper interaction in the exploited ecosystem of the northern continental shelf of Yucatan, Mexico. *Ecological Modelling* 129: 119–129.

Arreguín-Sánchez, F. 1996. Catchability: a key parameter for fish stock assessment. *Reviews in Fish Biology and Fisheries* 6 (2): 221–242.

———. 2012. The dynamics linking biological hierarchies, fish stocks and ecosystems: implications for fisheries management. In *Models of the ecological hierarchy: from molecules to the ecosphere*, ed. F. Jordán and S.E. Jørgensen, 501–516. Philadelphia, PA: Elsevier B.V.

Arreguín-Sánchez, F., and T.J. Pitcher. 1999. Catchability estimates and their application to the red grouper (*Epinephelus morio*) fishery of the Campeche Bank, Mexico. *Fishery Bulletin* 97: 746–757.

Arreguín-Sánchez F., P. del Monte-Luna, M.J., and Zetina-Rejón. 2015. Climate Change Effects on Aquatic Ecosystems and the Challenge for Fishery Management: Pink Shrimp of the Southern Gulf of Mexico. Fisheries 40(1): 15–19.

Arreguín-Sánchez F., P. del Monte-Luna, M.O. Albañez-Lucero, M.J. Zetina-Rejón, A. Tripp-Quezada, T.M. Ruiz-Barreiro, and J.C. Hernández-Padilla. 2021. How Much Biomass Must Remain in the Sea After Fishing to Preserve Ecosystem Functioning? The Case of the Sardine Fishery in the Gulf of California, Mexico. 143–161pp. In: Ortiz, M., F. Jordán (eds.), Marine Coastal Ecosystems Modelling and Conservation; Springer Nature, Switzerland.

Baranov, F.I. 1918. On the question of the biological basis of fisheries. *Nauchn Issled Ikhtiologicheskii Institute Izv* 1: 81–128.

Beverton, R.J.H., and S.J. Holt. 1957. On the dynamics of exploited fish populations. *Fisheries Investigations* 2 (19): 533.

Bojó, R.L. 1985. La Conferencia Mundial de la FAO (1) sobre ordenación y desarrollo pesqueros. *Boletín Económico de ICE, Información Comercial Española* 1992: 2065–2072.

Caddy, J.F. 1996. Modelling natural mortality with age in short-lived invertebrate populations: definition of a strategy of gnomonic time division. *Aquatic Living Resources* 9: 197–207.

Christensen, V., and C.J. Walters. 2005. Using ecosystem modeling for fisheries management: where are we? In *ICES Annual Science Conference, Aberdeen, Scotland, 20–24 September 2005*. ICES CM 2005/M:19.

Cicin-Sain, B., and R.W. Knecht. 1989. *The emergence of a regional ocean regime in the south Pacific*, Working paper no. 14. Honolulu: Environment and Policy Institute, East-West Center.

Clark, C.W., and G.R. Munro. 2002. The problem of overcapacity. *Bulletin of Marine Science* 70 (2): 473–483.

Coll, M., L.J. Shannon, K.M. Kleisner, M.J. Juan-Jordá, A. Bundy, A.G. Akoglu, D. Banaru, J.L. Boldt, M.F. Borges, A. Cook, I. Diallo, C. Fu, C. Fox, D. Gascuel, L.J. Gurney, T. Hattab, J.J. Heymans, D. Jouffre, B.R. Knight, S. Kucukavsar, S.I. Large, C. Lynam, A. Machias, K.N. Marshall, H. Masski, H. Ojaveer, C. Piroddi, J. Tam, D. Thiao, M. Thiaw, M.A. Torres, M. Travers-Trolet, K. Tsagarakis, I. Tuck, G.I. van der Meeren, D. Yemane, S.G. Zador, and Y.-J. Shin. 2016. Ecological indicators to capture the effects of fishing on biodiversity and conservation status of marine ecosystems. *Ecological Indicators* 60: 947–962.

Dinga, Q., X. Chen, R. Hilborn, and Y. Chen. 2017. Vulnerability to impacts of climate change on marine fisheries and food security. *Marine Policy* 83: 55–61.

Doll, J.C., and S.J. Jacquemin. 2019. Bayesian model selection in fisheries management and ecology. *Journal of Fish and Wildlife Management* 10 (2): 691–707.

FAO. 1995. *Code of conduct for responsible fisheries*. Rome: FAO. 41 p.

———. 2020. *El estado mundial de la pesca y la acuicultura 2020. La sostenibilidad en acción*. Rome: FAO. https://doi.org/10.4060/ca9229es.

Fariñas-Franco, J.M., A.L. Allcock, and D. Roberts. 2018. Protection alone may not promote natural recovery of biogenic habitats of high biodiversity damaged by mobile fishing gears. *Marine Environmental Research* 135 (2018): 18–28.

Fox, W.W., Jr. 1970. An exponential surplus-yield model for optimizing exploited fish populations. *Transactions of the American Fisheries Society* 99 (1): 80–88.

Gaines, S.D., C. Costello, B. Owashi, T. Mangin, J. Bone, J. García Molinos, M. Burden, H. Dennis, B.S. Halpern, C.V. Kappel, K.M. Kleisner, and D. Ovando. 2018. Improved fisheries management could offset many negative effects of climate change. *Science Advances* 4: eaao1378.

Gulland, J.A. 1965. Estimation of mortality rates. Annex to Arctic Fisheries Working Group report (meeting in Hamburg, January 1965). In *ICES, C.M. 1965*. Doc. No. 3 (mimeographed).

———. 1987a. Natural mortality and size. *Marine Ecology Progress Series* 39: 197–199.

———. 1987b. The effect of fishing on community structure. *South African Journal of Marine Science* 5 (1): 839–849.

———. 1987c. Seals and fisheries: case for predator control? *Trends in Ecology & Evolution* 2 (4): 102–104.

Hannesson, R. 1983. Optimal harvesting of ecologically interdependent fish species. *Journal of Environmental Economics and Management* 10 (4): 329–345.

Hilborn, R. 2012. The evolution of quantitative marine fisheries management 1985-2010. *Natural Resources Modelling* 25 (1): 122–144.

Jennings, S., and M.J. Kaiser. 1998. The effects of fishing on marine ecosystems. *Advances in Marine Biology* 34: 201–352.

Jones, R. 1981. The use of length composition data in fish stock assessments (with notes on VPA and cohort analysis). In *FAO fisheries circular*, vol. 734. Rome: FAO. 55 p.

Lam, V.W.Y., E.H. Allison, J.D. Bell, J. Blythe, W.W.L. Cheung, T.L. Frölicher, M.A. Gasalla, and U.R. Sumaila. 2020. Climate change, tropical fisheries and prospects for sustainable development. *Nature Reviews* 1: 440. http://labpesq.io.usp.br/images/publicacoes/_Lam_et_al_2020.pdf.

Lele, S.R., B. Dennis, and F. Lutscher. 2007. Data cloning: easy maximum likelihood estimation for complex ecological models using Bayesian Markov chain Monte Carlo methods. *Ecology Letters* 10: 551–563.

Levy, M.J.P. 1993. United Nations Convention on the Law of the Sea (UNCLOS). In *Green globe yearbook: an independent publication on environment and development*, 148. Norway: The Fridtjof Nansen Institute.

Lewis, A., K. Barnett, E.J. Ward, J.E. Jannot, and A.O. Shelton. 2019. Dynamic spatial heterogeneity reveals interdependence of marine faunal density and fishery removals. *Ecological Indicators* 107: 105585.

Liang, C., and D. Pauly. 2017. Fisheries impacts on China's coastal ecosystems: unmasking a pervasive 'fishing down' effect. *PLoS One* 12 (3): e0173296.

———. 2020. Masking and unmasking fishing down effects: the Bohai Sea (China) as a case study. *Ocean and Coastal Management* 184: 105033.

Lluch-Cota, S.E., A. Parés-Sierra, V.O. Magaña-Rueda, F. Arreguín-Sánchez, G. Bazzino, H. Herrera-Cervantes, and D. Lluch-Belda. 2010. Changing climate in the Gulf of California. *Progress in Oceanography* 87: 114–136.

Lluch-Cota, S.E., F. Arreguín-Sánchez, C.J. Salvadeo, and P. Del Monte Luna. 2018. Climate change impacts, vulnerabilities and adaptations: north-east tropical Pacific marine fisheries. In *Impacts of climate change on fisheries and aquaculture: synthesis of current knowledge, adaptation and mitigation options*, FAO Fisheries and Aquaculture technical paper no. 627, ed. M. Barange, T. Bahri, M.C.M. Beveridge, K.L. Cochrane, S. Funge-Smith, and F. Poulain, 207–218. Rome: FAO. 628 p.

Martínez-Aguilar, S., F. Arreguín-Sánchez, and E. Morales-Bojórquez. 2005. Natural mortality and life history stage duration of Pacific sardine (Sardinops caeruleus) based on gnomonic time divisions. *Fisheries Research* 71: 103–114.

Mendenhall, E., C. Hendrix, E. Nyman, P.M. Roberts, J.R. Hoopes, J.R. Watson, V.W.Y. Lam, and U.R. Sumaila. 2020. Climate change increases the risk of fisheries conflict. *Marine Policy* 117: 103954.

Morishita, J. 2008. What is the ecosystem approach for fisheries management? *Marine Policy* 32 (1): 19–26.

Murawski, S.A. 1991. Can we manage our multispecies fisheries? *Fisheries* 16 (5): 5–13.

———. 1993. Climate change and marine fish distributions: forecasting from historical analogy. *Transactions of the American Fisheries Society* 122 (5): 647–658.

Ortuño-Crespo, G., and D.C. Dunn. 2017. A review of the impacts of fisheries on open-ocean ecosystems. *ICES Journal of Marine Science* 74 (9): 2283–2297.

Pauly, D., and G.R. Morgan. 1987. Length-based methods in fisheries research. In *ICLARM Conference Proceedings*, vol. 13. 468 p.

Pella, J.T., and P.K. Tomlinson. 1969. A generalized stock production model. *Bulletin - Inter-American Tropical Tuna Commission* 13 (3): 421–458.

Pérez-Pérez, M., M. Dolores-Garza-Gil, and M. Varela-Lafuente. 2017. Fisheries management under ecological interdependence: the case of European Hake and Blue Whiting Fishery. *Natural Resources* 8: 569–581.

Phillips, B.F., and M. Perez-Ramírez, eds. 2018. *Climate change impacts on fisheries and aquaculture*. New York, NY: Wiley & Sons. Volumes 1 and II.

Pope, J.G. 1972. An investigation of the accuracy of virtual population analysis using cohort analysis. In *International Commission for the Northwest Atlantic Fisheries, research bulletin*, 65–74.

Pope, J.G., and J.G. Shepherd. 1982. A simple method for the consistent interpretation of catch-at-age data. *ICES Journal of Marine Science* 40 (2): 176–184.

———. 1985. A comparison of the performance of various methods for tuning VPAs using effort data. *ICES Journal of Marine Science* 42 (2): 129–151.

Punt, A.E., and R. Hilborn. 1997. Fisheries stock assessment and decision analysis: the Bayesian approach. *Reviews in Fish Biology and Fisheries* 7: 35–63.

———. 2001. *BAYES-SA - Bayesian stock assessment methods in fisheries – user's manual*, FAO computerized information series (Fisheries). No. 12. Rome: FAO. 56 p.

Reykjavik Declaration. 2001. Reykjavik Declaration on responsible fisheries in the marine ecosystem. In *Conference on Responsible Fisheries in the Marine Ecosystem, Reykjavik, Iceland, 1–4 October 2001*. 4 p.

Richardson, K., D. Haynes, A. Talouli, and M. Donoghue. 2017. Marine pollution originating from purse seine and longline fishing vessel operations in the Western and Central Pacific Ocean, 2003–2015. *Ambio* 46: 190–200.

Richter, A., A.M. Eikeset, D. van Soest, F.K. Diekert, and N.C. Stenseth. 2018. Optimal management under institutional constraints: determining a total allowable catch for different fleet segments in the northeast arctic cod fishery. *Environmental and Resource Economics* 69: 811–835.

Schaefer, M.B. 1954. Some aspects of the dynamics of populations important to the management of commercial marine fisheries. *Bulletin Inter-American Tropical Tuna Commission* 1: 27–56.

Schmiing, M., H. Diogo, R.S. Santos, and P. Afonso. 2015. Marine conservation of multispecies and multi-use areas with various conservation objectives and targets. *ICES Journal of Marine Science* 72 (3): 851–862.

Shephard, S., Da.G. Reid, H.D. Gerritsen, and K.D. Farnsworth. 2015. Estimating biomass, fishing mortality, and "total allowable discards" for surveyed non-target fish. *ICES Journal of Marine Science* 72 (2): 458–466.

Sitarz, D. 1993. *Agenda 21: the earth summit strategy to save our planet*. Oak Ridge, TN: U.S. Department of Energy.

Smith, T.D. 1994. *Scaling fisheries, the science of measuring the effect of fishing 1855–1955*. Cambridge: Cambridge University Press.

Stewart, I.J., A.C. Hicks, I.G. Taylor, J.T. Thorson, C. Wetzel, and S. Kupschus. 2013. A comparison of stock assessment uncertainty estimates using maximum likelihood and Bayesian methods implemented with the same model framework. *Fisheries Research* 142: 37–46.

Torrejón-Magallanes, J., E. Morales-Bojórquez, and F. Arreguín-Sánchez. 2021. Improving the gnomonic approach with the gnomonicM R-package to estimate natural mortality throughout different life stages. *PeerJ* 9: e11229.

Tudela, S. 2004. *Ecosystem effects of fishing in the Mediterranean: an analysis of the major threats of fishing gear and practices to biodiversity and marine habitats. Studies and reviews*. Vol. 74. Rome: General Fisheries Commission for the Mediterranean. 44 p.

UN. 1992. *Agenda 21 United Nations Conference on Environment and Development, Rio de Janeiro. Brazil, 3–14 June 1992*. New York, NY: UN. http://www.un.org/esa/sustdev/documents/agenda21/index/htm.

———. 2002. *World Summit on sustainable development. The Johannesburg plan of implementation*. New York, NY: United Nations. http://www.johannesburg.org.

Verhulst, P.F. 1838. Notice sur la loi que la population poursuit dans son accroissement. *Correspondance Mathématique et Physique* 10: 113–121.

Xing, L., Y. Chen, R. Boenish, K.R. Tanaka, N. Barrier, and Y. Ren. 2021. Evaluating the impacts of fishing and migratory species in a temperate bay of China using the ecosystem model OSMOSE-JZB. *Fisheries Research* 243: 106051.

Young, O.R. 1982. Regime dynamics: the rise and fall of international regimes. *International Organization* 36 (2): 277–297.

Zeller, D., and D. Pauly. 2019. Viewpoint: back to the future for fisheries, where will we choose to go? *Global Sustainability* 2 (e11): 1–8.

Chapter 3
Ecosystem Trophic Structure and Dynamics: Basic Concepts

Theoretical Basis of Ecosystem Dynamics

Food webs are the set of interdependencies between organisms through food relationships. Who eats who establishes a relationship of survival between organisms, and food arrangements lead to the structure, organization, and functioning of communities and ecosystems. If we imagine these relationships as a set of nodes, representing the species, and the links between them, that is, who relates to whom, trophically, two elements inherent to trophic networks arise: topological structure, that is, who connects with who, and its functioning, that is, the energy flow between species through what they consume. Eventually, when the intensities of these flows are varied, for example, by food preferences, by prey vulnerability, or by consumption rates, then the cohesion between the network elements varies as well, lending them organizational properties. In this way, the study of the dynamics of ecosystems, represented by connections, the energy flow between them, and the variations over time, will allow an understanding of the dynamics of ecosystems and will define attributes and indicators that may eventually be useful for conservation and exploitation.

Imagine a trophic network, where nodes represent species and connections represent trophic relationships between species. Let us define i as the prey or donor and j as the predator or receiver. In terms of flow T_{ij} represents the joint probability that the events i and j co-occur[1] and represents the probability that an energy particle will flow from i a j. According to Ulanowicz (1986, 2009) and following the basic concepts of information theory of Boltzmann (1872), Shannon and Weaver (1949), and MacArthur (1955), the probability of non-occurrence of that combination of events is represented as $s_i = \log(1 - T_{ij}) = -k \log(T_{ij})$, that is, the probability that the particle coming from i reaches any other species except j. The

[1] The probability is given by the proportion of flows from i to j relative to the total flows; that is, $\frac{T_{ij}}{T_{\circ,\circ}}$.

F. Arreguín-Sánchez, *Holistic Approach to Ecosystem-Based Fisheries Management*, https://doi.org/10.1007/978-3-030-96847-2_3

product of these two probabilities, $s_{ij} = T_{ij}s_i = kT_{ij} \log (T_{ij})$, is known as *indeterminacy* (Ulanowicz et al. 2009) and refers to a property defined by the fact that for a moving particle, and it is not possible to simultaneously determine velocity and position; hence, it is an intrinsic attribute of a dynamic system. In these relations, k represents a scalar constant, and Ulanowicz (1986) assigns the value of the total flows to reflect a dimension of size of the system.

Following Ulanowicz et al. (2009), if the events i and j are independent, the marginal probability that i occurs is $T_{i\circ} = \sum_j T_{ij}$, that is, the flows of a particular prey i towards all predators, while the probability that j occurs independently of i is $T_{\circ j} = \sum_i T_{ij}$, representing the flows from all prey to a particular predator, where the symbol $\circ$ represents the sum of the prey or predator as appropriate to their position in the notation. In this way, it is assumed that indeterminacy is maximum when i and j are completely independent, represented as s^*_{ij}; the difference $s^*_{ij} - s_{ij}$ represents the extent of the limitation or conditions of flows that i exerts on j, represented as $x_{i\,|\,j}$, such that:

$$x_{i|j} = s^*_{ij} - s_{ij} = -k \cdot \log\left(T_{i\circ} T_{\circ j}\right) - \left[-k \cdot \log\left(T_{ij}\right)\right] = k \log\left(\frac{T_{ij}}{T_{i\circ} T_{\circ j}}\right) = x_{j|i}$$

which means that the expression also represents the limitation or conditioning that j exerts on i; that is, the expression represents a mutual limitation. In this context, the limitation refers to the maintenance of the flow between i and j, but not to the possible variations in its magnitude.

In this way, the average mutual limitation for all flows in the food web would be the sum of each flow weighted by the limitation of each interaction, such that

$$X = k\sum_{i,j} T_{ij} \log\left(\frac{T_{ij}}{T_{i\circ} T_{\circ j}}\right)$$

Note that the limitation of flows is associated, topologically, with the way in which the connectivity of the network is organized and with the magnitude of the flows, being, consequently, an organizational attribute of the ecosystem.

According to Ulanowicz (1986) and Ulanowicz et al. (2009), aggregate indeterminacy over the entire trophic network is the upper limit of how much limitation (organization) the trophic web can contain. Thus, if this upper limit is represented as H, then

$$H = -k\sum_{i,j} \frac{T_{ij}}{T_{\circ\circ}} \log\left(\frac{T_{ij}}{T_{\circ\circ}}\right)$$

and

$$\mathcal{E} = H - X = -k\sum_{i,j} T_{ij} \log\left(\frac{T_{ij}^2}{T_{i\circ}T_{\circ j}}\right)$$

where $\mathcal{E}$ is conditional entropy.

As mentioned before, Ulanowicz (1986) assigns to constant k the dimension $T_{\circ\circ}$ to scale the different estimators according to the size of the system that is measured by the total system throughput and redefines the previous relationships as:

$$A = \sum_{i,j} T_{ij} \log\left(\frac{T_{ij}T_{\circ\circ}}{T_{i\circ}T_{\circ j}}\right) \tag{3.1}$$

defined as ascendency,

$$C = -\sum_{i,j} T_{ij} \log\left(\frac{T_{ij}}{T_{\circ\circ}}\right) \tag{3.2}$$

defined as development capacity, and

$$\phi = -\sum_{i,j} T_{ij} \log\left(\frac{T_{ij}^2}{T_{i\circ}T_{\circ j}}\right) \tag{3.3}$$

defining conditional entropy as the reserve energy or development capacity of the system,

$$\text{where } C = A + \phi$$

Ecosystem Organization, a Key Concept

Ulanowicz et al. (2009) defines $\frac{A}{C} = a$ as a relative measure of the organization of the ecosystem, that is, of the power or organizational capacity that flows in the system, with $1 > a > 0$. Using the same formulation related to indeterminacy, the function $F = -ka \log(a)$ is defined as the ability of the system to change or self-organize and modifies the relationship in such a way that it reaches a maximum. To do this, a fit parameter is introduced, β, and the constant k is normalized, such that $F_{\max} = 1$; thus

$$F = -\left[\frac{e}{\log(e)}\right] a^{\beta} \log\left(a^{\beta}\right) \tag{3.4}$$

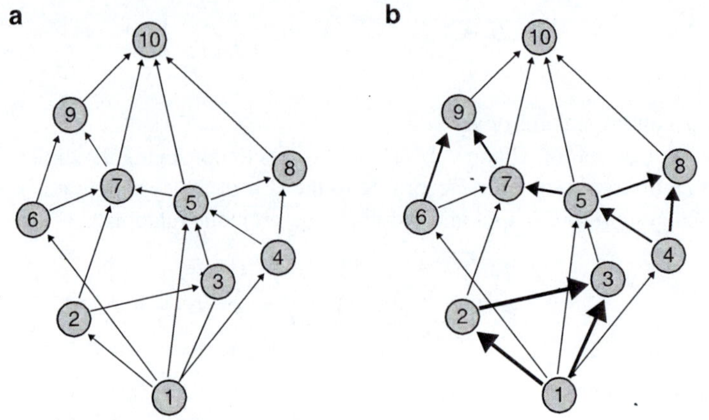

Fig. 3.1 Schematic representation of two trophic networks, in which the intensity of flows is similar between the components (**a**) and where the intensity of flows between nodes are different (**b**) and exemplified by arrow thickness

which represents the fraction of the activity that is effectively used by the system to maintain a balance between A and ϕ. In this way, in absolute terms, the robustness of the system can be calculated as $R = T_{\circ\circ} \cdot F$, and, dynamically, the change in the robustness of the system due to a change in flow can be given as:

$$\frac{\partial R}{\partial T_{ij}} = F + \frac{T_{\circ\circ}F'}{C}\left\{ \log\left[\frac{T_{ij}T_{\circ\circ}}{T_{i\circ}T_{\circ j}}\right] + a\log\left[\frac{T_{ij}^2}{T_{i\circ}T_{\circ j}}\right]\right\} \tag{3.5}$$

where F' is the derivative with respect to a

$$F' = +e\beta a^{\beta-1}\left[\frac{\log\left(a^{\beta}\right)}{\log(e)} + 1\right]$$

In this way, when the system is optimal, $F = 1$, and $F' = 0$, and when $a < a_{\text{opt}}$, F' will be positive (the sum of inflows and outflows of a network compartment will be positive).

An aspect of great interest for understanding the dynamics of ecosystems through attributes such as measures of organization $\left(\frac{A}{C}\right)$, capacity for self-organization (F), development capacity (ϕ), and robustness (R) is to clearly identify the role of energy flows between ecosystem components. Let us analyze the two trophic networks in Fig. 3.1. The networks have the same topology (nodes and connections) but differ in the magnitude of the flows, indicated by the thickness of the arrows, which indicate the direction of the flows, presumably starting at node 1, which represents the primary producers, and progressing to node 10, which represents the top predators. In network a, the intensity of the flows is equal between the nodes, such that

$T_{1,\,3} = T_{4,\,5} = T_{7,\,10}$, etc.; and for network *b*, shown on the right, $T_{1,\,3} > T_{4,\,5} > T_{7,\,10}$.

In the case of ascendency, for example, if we omit the sum, we have: $A_{i,j} = T_{i,j} \log\left(\frac{T_{ij}T_{\circ\circ}}{T_{i\circ}T_{\circ j}}\right)$, which represents the partial contribution to the total ascendency of the flow between *i* and *j*. In trophic network *a* in Fig. 3.1, the partial ascendency between nodes is equal, that is, $A_{1,\,3} = A_{4,\,5} = A_{7,\,10}$. Thus, for example, if the biomass of a node is reduced, and the intensity of the flow between predator and prey changes, the partial ascendency changes, but in terms of the system, the effect would be the same regardless of which node of the system was affected. The same result would occur with the capacity for self-organization, *F*, or with the robustness indicator, *R*. On the other hand, if we imagine that, in Fig. 3.1, the thickness of the arrows of network *b* represents the partial magnitude of the flow between nodes, then $A_{1,\,3} \neq A_{4,\,5} \neq A_{7,\,10}$. For example:

$$A_{1,3} = T_{1,3} \log\left(\frac{T_{1,3}T_{\circ\circ}}{T_{1,\circ}T_{\circ,3}}\right) > A_{7,10} = T_{7,10} \log\left(\frac{T_{7,10}T_{\circ\circ}}{T_{7,\circ}T_{\circ,10}}\right)$$

In this way, the effect of a change in the biomass of node 1 will not be similar to the change in biomass of node 7, even when the change is similar and both act as donors or prey; equally, the effect would not be similar to the capacity for self-organization, *F*, and robustness, *R*, because the partial contribution of the flows between nodes in the trophic network is different. This notion is relevant because in nature, the interaction among organisms in a trophic network is as exemplified in network *b* (Fig. 3.1), and this knowledge is key when trying to assess the effects on ecosystems of biomass loss, like fishing.

From the topological point of view, that is, nodes and connections, the literature defines different attributes according to some of the inherent properties, such as centrality indices (Freeman 1978), that identify different attributes of the network according to their connectivity and based on the quantification of the number of connections, the participation of each group in possible flow pathways, the differential capacity between species to connect with others in the trophic network, and the frequency with which a species appears in trophic pathways, among others. Some of these attributes are used to identify key species (Libralato et al. 2006; Jordán et al. 2006). These topological indicators are useful for identifying the role of species in ecosystems, allowing, in turn, the determination of which ones play a preponderant role in certain aspects or are critical for the degree of connectedness in the network, that is, their participation in connectivity or, in contrast, dysconnectivity. In Table 3.1, some of the topological indicators used for network analyses are summarized.

If, in addition to the topological attributes of the network, the energy flow inherent to each interaction is considered, analyses of system topology usually provide information of great relevance. Thus, for example, if network *a* in Fig. 3.1 is examined, there are 17 different pathways through which a particle can get from node 1 to 10; if node 5 is eliminated, the number of pathways is reduced to 6, but if node 9 is eliminated, the number of pathways is reduced to 13. This essentially

Table 3.1 Topological indicators used for trophic network analysis

Index name	Formula	Interpretation	References
Node degree	$C_D(n_i) = \sum_{j=1}^{s} x_{ij} = \sum_{j=1}^{s} x_{ji}$	Sum of number of prey and predators	Fedor and Vasas (2009)
Connectance	$C = \frac{2L}{N(N-1)}$	The ratio of realized and possible links in the network	Jordán and Scheuring (2004)
Weighted connectance	$C_w = \frac{LD_w}{N^2} LD_W = e^{0.5\phi}$	Incorporates the strength of trophic links	Boit and Gaedke (2014)
Average path length	$APL_w = \frac{1}{N(N-1)} \sum_{ij}^{N} d_{ij}$	The mean length of the shortest paths for all possible pairs of nodes in the network	Oshima and Leaf (2018)
Betweenness centrality	$BC_i = \frac{\sum_{j<k} g_{jk}(i)/g_{jk}}{(1/2)(N-1)(N-2)}$	How frequently a node i is on the shortest paths between every pair of nodes j and k	Fedor and Vasas (2009)
Closeness centrality	$CC_i = \frac{N-1}{\sum_{j=1}^{N} d_{ij}}$	Minimum number of links from an individual node to all others in the network	Wassermann and Faust (1994)
Fragmentation measure	$F_i = 1 - \frac{\sum_k S_k(S_k-1)}{N(N-1)}$	The importance of a given node based on network fragmentation after its deletion	Fedor and Vasas (2009)
Distance-weighted reach	$RD_i = \frac{\sum_j \frac{1}{d_{ij}}}{N}$	The sum of the reciprocals of distances from node i to all nodes	Fedor and Vasas (2009)
Topological importance	$TI_i^n = \frac{\sum_{m=1}^{n} \sum_{j=1}^{N} a_{m,ij}}{n} \ldots a_{l,ij} = \frac{1}{ND_j}$	Is for characterizing long indirect effects. The effects of i on j in m steps is given as $a_{m,ij}$	Jordán et al. (2006)
Keystoneness index	$K_i = \sum_{j=1}^{n} \frac{1}{d_{ij}}(1 + K_{bj}) + \sum_{e=1}^{m} \frac{1}{f_e}(1 + K_{le})$	Emphasizes vertical (bottom-up and top-down) over horizontal interactions	Jordán et al. (1999)
Modularity	$M_W = \sum_{S=1}^{N_M} \left[\left(\frac{W_s^{in}}{W} \right) - \left(\frac{W_s^{all}}{2W} \right)^2 \right]$	Identifies communities in the food web under the concept that connections are more intense within the community than between them	Nascimento and Pitsoulis (2013), Newman and Girvan (2004)
Compartmentation of the network	$C_1 = \frac{\sum_{ij} p_{ij}}{N(N-1)}$	Quantifies whether there exist groups of species with many interactions within the group but only a few interactions between members of different groups	Jordán and Scheuring (2004)

Average distance	$d_{av} = \frac{2\sum_{ij} d_{ij}}{N(N-1)}$	Average distance between pairs of nodes, i.e., the distance between two random nodes	Jordán and Scheuring (2004)
Omnivory index	$\mathrm{OI}_j = \sum_{i=1}^{S} (\mathrm{TL}_i - \mathrm{TL})^2 \mathrm{DC}_{ij}$	Omnivory of species	Jordán and Scheuring (2004)
Network flow reliability	$R = 1 - \sum_{i=1}^{N-1} \binom{N-1}{i} p^i q^{(N-1-i)} F_i$	The probability that the single top-predator will be unconnected to any of the sources after deleting other nodes (except the sink node)	Jordán and Scheuring (2004)
K-Player-1 problem	$F = 1 - \frac{\sum_k (S_k - 1)}{N(N-1)}$	Deletion results in the fragmentation of the original network, consider the distances between nodes	Borgatti (2003), Jordán (2005)
K-Player-2 problem	$F^{\mathrm{D}} = 1 - \frac{2\sum_{i>j} \frac{1}{d_{ij}}}{N(N-1)}$	Deletion results in the fragmentation of the original network, considers reciprocal distances (otherwise, disconnected nodes are infinitely distant from each other)	Borgatti (2003), Jordán (2005)
K-Player-2 problem	$R^{\mathrm{D}} = \frac{\sum_j \frac{1}{d_{sj}}}{N}$	Quantifying how easily (shortly) other nodes are reachable from the focal one	Borgatti (2003), Jordán (2005)
Effective number of flows	$f = \prod_{ij} \left(\frac{1}{t_{ij}}\right)^{t_{ij}}$	Effective trophic role	Ulanowicz et al. (2014)
Effective number of roles	$r = \prod_{ij} \left(\frac{t_{ij}}{t_{i\circ} t_{\circ j}}\right)^{t_{ij}}$	Measure of how many distinct functions are in the network (number of effective trophic levels)	Ulanowicz et al. (2014)
Link density	$c = \prod_{ij} \left(\frac{\sqrt{t_{i\circ} t_{\circ j}}}{t_{ij}}\right)^{t_{ij}}$	Weighted average of number of inflows and outflows of a node (considers number of flows and their magnitude)	Ulanowicz et al. (2014)
Degree of organization	$a = \frac{\ln(r)}{\ln(r) + 2\ln(c)}$	Represents an empirical constraint between the effective number of trophioc levels (roles, r) and link density (c), where the upper (and lower) limit of one implies the upper (or lower) limit of the other	Ulanowicz et al. (2014)
Link density	$c = r^{(1-a)/(2a)}$	link density (c) is related with effective number of roles(r) and the degree of organization of the network (a)	Ulanowicz et al. (2014)

shows that each node has a different contribution to the system topology. If we now observe network *b*, with different flow intensities between nodes, it is evident that if node 5 is eliminated, the amount of energy that would potentially reach node 3 would not be "used" by other species, not providing a significant amount of energy to nodes 7 and 8 (judging by the thickness of the arrows), resulting in dissipated energy (assuming that the alternative pathways do not change significantly with respect to their flow intensities). Of course, this type of process has repercussions throughout the system because the signal or impact, given by the magnitudes of energy that flow or stop flowing in the system, propagates throughout the entire trophic network; that is, there are direct and indirect impacts, as exemplified by the matrix of Leontief (1951).

On the other hand, there is extensive literature that refers to the internal structure of trophic networks, in our case representing ecosystems and of particular interest are those attributes that are defined as modularity, a concept that refers to the formation of subsets, within the trophic network, whose elements (species) interact differently with them than with other elements in the network (Seidman and Foster 1978; Borgatti 2003; Borgatti et al. 1990; Girvan and Newman 2002) and, as such, have a differentiated dynamic from other subsets. For example, if the intensity of network flows is greater or less among a set of nodes with respect to other nodes in the network, there will be differences in cohesion, a relevant aspect for the maintenance of trophic network organization. In nature, subsets are associated with habitat properties, including spatial dimensions, life history, diet preferences, and the size of organisms, and, in general, they change and are determined by responding to the adaptability of the species throughout evolution (Pimm and Lawton 1980; Allesina et al. 2009; Rezende et al. 2009; Caddy and Stamatopoulos 1990; Caddy 2014). In general, it is assumed that the existence of subsets is associated with cohesion attributes of trophic networks, which is related to dynamic stability and persistence (Stouffer and Bascompte 2011; Saavedra et al. 2011).

To better understand these concepts, let us return again to the example of trophic network *b* in Fig. 3.1. The module or subset composed of nodes 1–2–3 clearly shows greater interactions among them than with other elements or subsets in the network, judging by the intensity of the interactions; the same occurs with nodes 4–5–8, although their interactions are less intense than the previous subset but greater than the existing interactions with other elements. Let us consider the flow $T_{2,\ 3}$ and Eq. (3.3); clearly, this flow contributes more to energy in reserve than does, for example, the flows $T_{4,\ 5}$ or $T_{1,\ 6}$. This same concept is applied to ecosystem resilience estimators, defined by Arreguín-Sánchez (2014) as $\mathrm{Res} = -\log\left(\frac{T_{ij}^2}{T_{i\circ}T_{\circ j}}\right)$, interpreted under the same rationale as that for Eq. (3.3). This definition of resilience, closely associated with the energy in reserve expressed by Eq. (3.3), also illustrates how the intensity of flows among elements of an ecosystem and within the functional structure can contribute to the vulnerability of ecosystems; for example, with less resilience or less energy in reserve or with less capacity for self-organization, ecosystems would be more vulnerable. In this context, it is evident that the excessive extraction of biomass from an ecosystem will strongly reduce not only the robustness of the fish resource but also the ability of the ecosystem to recover.

Ecosystem Metabolism: Supply–Demand of Energy

On the other hand, some authors have established the existence, in nature, of a consistent ratio between the baseline metabolism of species and their body mass that essentially reflects the efficiency of the individual transport network (i.e., metabolites in the blood); this principle is known as the ¾ power allometric scaling law (Kleiber 1932; Hemmingsen 1960). This principle has been adapted to other types of networks in which the topology, structure, and size of the network determine the transport efficiency of the system (West et al. 1997; West and Brown 2005; Banavar et al. 2002, 2010). Bendoricchio and Palmeri (2005) refer to this scaling exponent as an index that determines the state of the ecosystem according to the balance between the supply and demand (SDB) of energy, by establishing that the relationship between the flow and biomass of compartments is analogous to the allometric scaling exponent (¾) of the species. In addition, these authors suggest that this slope is an indicator of changes in the ecosystem, where values greater than ¾ indicate a situation in which the supply exceeds the demand for energy, and the ecosystem is consequently more resilient. Finally, these authors suggest a dynamic behavior of the balance between energy supply and demand that varies according to disturbances in an ecosystem. In this sense, the supply–demand balance of energy in an ecosystem is affected by the dynamics of its components and the system as a whole and is closely linked to the redistribution of biomass and energy flows. Salcido-Guevara and Arreguín-Sánchez (2014) demonstrated the regularity of the concept of potential scaling of ¾ in marine ecosystems for a set of 98 ecosystems around the world, with no significant differences with latitude or by ecosystem type. An additional aspect of great interest is that the demonstrated allometric regularity is strongly associated with energy transfer efficiency.

One of the relevant aspects of the regularity of this allometric scaling indicator is that it reflects some attributes of interest for the identification of effects of impacts on an ecosystem. For example, Bendoricchio and Palmeri (2005) suggest that SDB values >0.75 correspond to trophic networks with a supply–demand ratio higher than the unit, and as a consequence, with high resilience. Power (1992) recognizes that the regularity of the SDB indicator implies, in turn, the limitations of flows in different trophic levels and the relationship of the indicator with bottom-up and top-down flow controls (sensu Huxel and McCann 1998). Additionally, Polis and Strong (1996) suggest that the dynamics of a system is primarily governed by the availability of resources (bottom-up control), although other studies highlight the role of flow control by predators (top-down control) (Hairston et al. 1960; Oksanen 1991; Huxel and McCann 1998). In this sense, it is easy to understand that the SDB indicator is an estimator that reflects possible alterations in the thermodynamic balance of an ecosystem. This aspect is of great interest for studies on sustainability. However, there are still specific aspects of processes linked to changes in the SDB index that need to be explored. Figure 3.2 presents the results of a simulation experiment in a benthic ecosystem in the Gulf of California (Arreguín-Sánchez et al. 2002) where shrimp biomass is extracted with a gradually increasing harvest

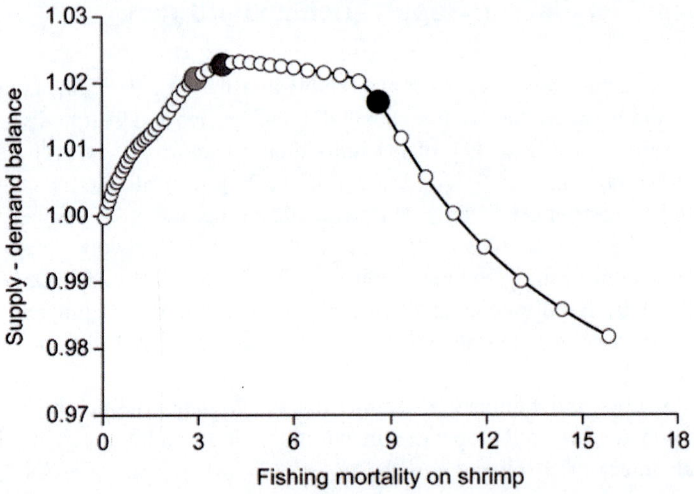

Fig. 3.2 Changes in the supply–demand balance of energy in the ecosystem of the central Gulf of California along the exploitation state of the shrimp stock. Gray point on left represents marginal yield; dark point at central region represents the state at the maximum sustainable yield, and the black point on the right side represents the reference limit after which the balance in supply–demand of energy falls abruptly, indicating the collapse of the shrimp fishery. The region between the MSY and limit reference point represents overfishing of the shrimp stock

rate. The trajectory of the supply–demand balance index of energy for the ecosystem indicates maximum values corresponding to the maximum sustainable yield level (Schaefer 1954) as well as the limit and threshold harvest level, after which the ecosystem collapses, corresponding to a harvest rate (capture/available biomass ratio) of 73%.

Populations and Ecosystem Performance

Another important aspect of ecosystem dynamic is the role that populations can play. For example, recently, there has been great interest in determining fishing limits for forage species that are often key to maintain the ecosystem functioning; or species that play a fundamental role in the biodiversity conservation. Also, the role of the species reflected in ecosystem attributes, such as resilience, vulnerability, functional impact, trophic replacement, and loss of production, among others (Jennings and Polunin 1996; Cury et al. 2005; Cury and Christensen 2005; Jennings 2005; Coll et al. 2008). In a practical sense, the dynamic relationship between individual populations and the ecosystem allows for a better understanding of their performance, and for obtaining greater knowledge for the sustainable use of these resources. To demonstrate this concept, let us assume that the performance of a population can be measured by its survival, which is expressed as (Allen 1971; Gulland and Garcia 1984):

$$S_i = \exp^{-Z_i} = \exp^{-(P/B)_i} \tag{3.6}$$

where Z is the instantaneous total mortality rate and P/B is the production/biomass ratio; which can be expressed in terms of flows as:

$$\left(\frac{P}{B}\right)_i = \sum_{j=1}^{n} \left(\frac{T_{i,j}}{T_{j,\circ}}\right)$$

Similarly, according to Zorach and Ulanowicz (2003), the role, ϱ_i, interpreted here as the performance of a population in an ecosystem can be represented as:

$$\varrho_i = \exp^{\mathrm{AMI}_i} \tag{3.7}$$

where AMI is the average mutual information (Ulanowicz 1986) and is expressed as:

$$\mathrm{AMI}_i = \sum_{i,j} \left(\frac{T_{i,j}}{T_{\circ,\circ}}\right) \log \left(\frac{T_{i,j} T_{\circ,\circ}}{T_{i,\circ} T_{\circ,j}}\right) \tag{3.8}$$

Figure 3.3 shows graphs of ϱ_i vs. S_i illustrating the annual trends for each functional group for the Atasta region model, in the southern Gulf of Mexico (Arreguín-Sánchez 2012). This figure was constructed through a simulation experiment of the effect of climate change represented by the time series of the SST over the last 60 decades.

The slope of the tendencies for each functional group is represented as:

$$\mathrm{slope}_i = \frac{\Delta \varrho_i}{\Delta S_i} = \frac{\Delta \exp^{\mathrm{AMI}_i}}{\Delta \exp^{(P/B)_i}}$$

which, when represented in terms of flows, and after a little algebra, becomes

$$\mathrm{slope}_i = \sum_{i,j} \left(\frac{[T_{\circ,\circ}]^2}{T_{i,\circ} T_{\circ,j}}\right) \tag{3.9}$$

Now, as mentioned above, according to Ulanowicz (1986), the ratio between ascendency and the development capacity, $\frac{A}{C}$, is a measure of organization; then, taking Eqs. (3.1) and (3.2):

$$\frac{A}{C} = \frac{\sum_{i,j} T_{ij} \log \left(\frac{T_{ij} T_{\circ\circ}}{T_{i\circ} T_{\circ j}}\right)}{\sum_{i,j} T_{ij} \log \left(\frac{T_{ij}}{T_{\circ\circ}}\right)} = \sum_{i,j} \left(\frac{[T_{\circ,\circ}]^2}{T_{i,\circ} T_{\circ,j}}\right) \tag{3.10}$$

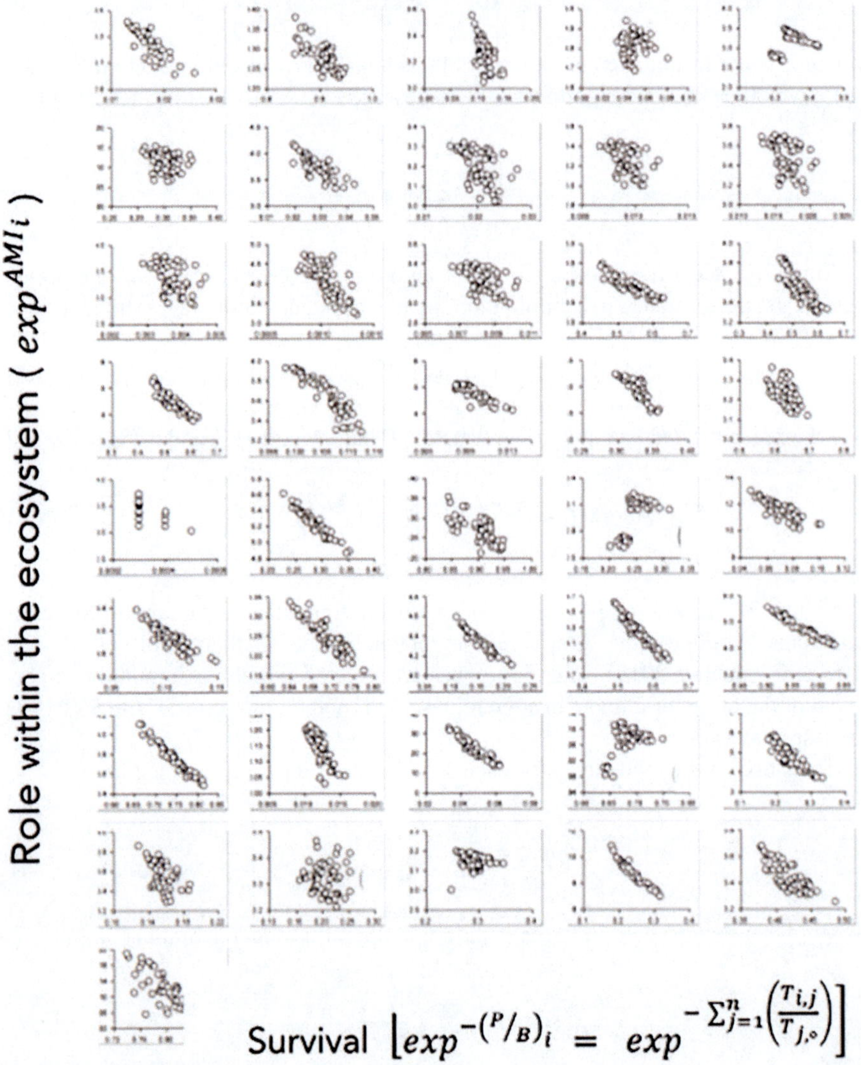

Fig. 3.3 Graph showing the performance trajectory of the functional groups of the coastal ecosystem of Atasta, southern Gulf of Mexico, because of climate change. Each cell represents a functional group of the ecosystem; and dots represents a year within the period 1960–2010. Cells represent, from top to down and left to right, the sequence of higher to lower trophic levels. The stock performance is given by its survival (*X*-axes) and by its trophic role in the ecosystem (*Y*-axes). All tendencies are negative, representing an ecosystem dominated by bottom-up control, since primary production is strongly limited as ecosystem resource (see Arreguín-Sánchez 2012; Arreguín-Sánchez et al. 2008, 2015, 2019)

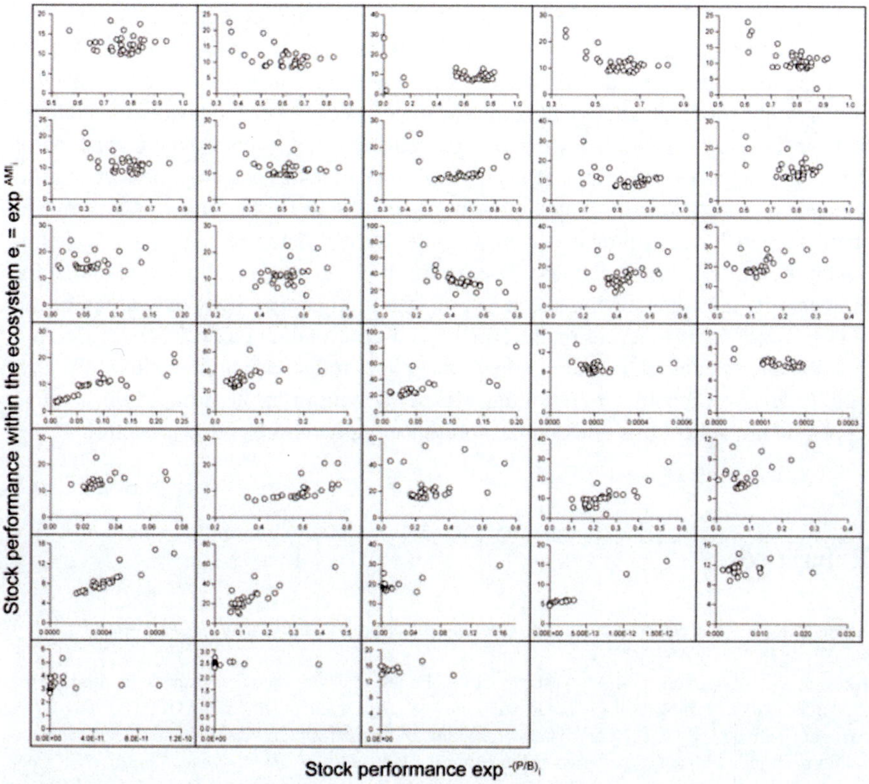

Fig. 3.4 Graph showing the performance trajectory of the functional groups of the northern Gulf of California ecosystem, because of climate change. Each cell represents a functional group of the ecosystem; and dots represent a year within the period 1960–2010. Cells from top to down and left to right represent the sequence of higher to lower trophic levels. The stock performance is given by its survival (*X*-axes) and by its trophic role in the ecosystem (*Y*-axes). Tendencies for low trophic levels are positive, while for top predator tend to be moderately negatives. Intermediate trophic level does not show tendency. In the northern Gulf of California, there are seasonal upwellings that promote primary production which, in turn, is not fully used in the system and is exported to neighbor ecosystems (see Arreguín-Sánchez et al. 2017b)

The expressions in Eqs. (3.9) and (3.10) are similar, with the exception that the second refers to the entire ecosystem and the first refers to a specific functional group. According to this, the slope of each functional group in the ecosystem, in Fig. 3.3, is interpreted as a relative measure of the contribution of each functional group to the organization of the ecosystem.

Figure 3.4 presents a similar simulation for an ecosystem in the northern region of the Gulf of California that incorporates the forcing effect of sea surface temperature anomalies on primary producers and couples the historical series from 1900 to date and projections to 2100 according to the IPCC (Collins et al. 2013).

Unlike Fig. 3.3, Fig. 3.4 shows that for lower trophic levels, at the bottom, the slopes of the trajectories are positive, while for the upper trophic levels, corresponding to top predators, the slopes are negative. In the intermediate trophic level groups, the slope is close to zero. The difference between the two ecosystems is that for the southern Gulf of Mexico, a decreasing trend in the carrying capacity of the ecosystem as a result of the effect of climate change has been reported (Arreguín-Sánchez et al. 2017a); in contrast, the northern region of the Gulf of California is characterized by the influence of upwellings during much of the year (Arreguín-Sánchez et al. 2017b). In the first case, primary production is a limited resource for the ecosystem, while for the second case, it is not. In this way, the slopes of the ϱ_i vs. S_i ratios seem to represent the type of flow control defined by the energy supply and demand of ecosystems; that is, negative slopes reflect constrained flow from low trophic levels, probably representing a type of bottom-up control, while positive slopes do not reflect this constraint, representing a type of top-down control.

References

Allen, K.R. 1971. Relation between production and biomass. *Journal of the Fisheries Board of Canada* 28 (10): 1573–1581.

Allesina, S., A. Bodini, and M. Pascual. 2009. Functional links and robustness in food webs. *Philosophical Transactions of the Royal Society B: Biological Sciences* 364 (1524): 1701–1709.

Arreguín-Sánchez, F. 2012. The dynamics linking biological hierarchies, fish stocks and ecosystems: implications for fisheries management. In *Models of the ecological hierarchy: from molecules to the ecosphere*, ed. F. Jordán and S.E. Jørgensen, 501–516. Oxford, UK: Elsevier B.V.

———. 2014. Adaptabilidad y sostenibilidad: un enfoque holístico hacia el manejo de ecosistemas acuáticos explotados. *Ciencia Pesquera* 22 (1): 89–101.

Arreguín-Sánchez, F., E. Arcos, and E.A. Chávez. 2002. Flows of biomass and structure in an exploited benthic ecosystem in the Gulf of California, Mexico. *Ecological Modeling* 156: 167–183.

Arreguín-Sánchez, F., P. del Monte Luna, M.J. Zetina-Rejón, A. Tripp-Valdez, M.O. Albañez-Lucero, and T.M. Ruiz-Barreiro. 2017a. Building an ecosystems-type fisheries management approach for the Campeche Bank, subarea in the Gulf of Mexico Large Marine Ecosystem. *Environmental Development* 22: 143–149.

Arreguín-Sánchez, F., P. del Monte-Luna, M.J. Zetina-Rejón, and M.O. Albáñez-Lucero. 2017b. The Gulf of California large marine ecosystem: fisheries and other natural resources. *Environmental Development* 22: 71–77.

Arreguín-Sánchez F., P. del Monte-Luna, and M.J. Zetina-Rejón. 2015. Climate Change Effects on Aquatic Ecosystems and the Challenge for Fishery Management: Pink Shrimp of the Southern Gulf of Mexico. *Fisheries* 40(1): 15–19.

Arreguín-Sánchez F., M. Ramírez-Rodríguez, M.J. Zetina-Rejón, and V.H. Cruz-Escalona. 2008. Natural hazards, stock depletion, and stock management in the southern Gulf of Mexico pink shrimp fishery. *American Fisheries Society Symposium* 64: 419–428.

Arreguín-Sánchez F., M.O. Albañez-Lucero, P. Del Monte-Luna, and M-J. Zetina-Rejón. 2019. Fishery Resource Management Challenges Facing Climate Change. 181–194pp. In: Ibañez A.L. (ed.). Mexican Aquatic Environments, Springer.

Banavar, J.R., J. Damuth, A. Maritan, and A. Rinaldo. 2002. Supply–demand balance and metabolic scaling. *Proceedings of the National Academy of Sciences of the United States of America* 99 (16): 10506–10509.

Banavar, J.R., M.E. Moses, J.H. Brown, J. Damuth, A. Rinaldo, R.M. Sibly, and A. Maritan. 2010. A general basis for quarter-power scaling in animals. *Proceedings of the National Academy of Sciences* 107 (36): 15816–15820.

Bendoricchio, G., and L. Palmeri. 2005. Quo vadis ecosystem? *Ecological Modelling* 184: 5–17.

Boit, A., and U. Gaedke. 2014. Benchmarking successional progress in a quantitative food web. *PLoS One* 9 (2): e90404.

Boltzmann, L. 1872. Weitere Studien über das Wärmegleichgewicht unter Gasmolekülen. *Wien Berlin* 66: 275–370.

Borgatti, S.P. 2003. The Key Player problem. In *Dynamic social network modeling and analysis: workshop summary and papers*, ed. R. Breiger, K. Carley, and P. Pattison, 241–252. Washington, DC: Committee on Human Factors, National Research Council.

Borgatti, S.P., M.G. Everett, and P.R. Shirey. 1990. LS sets, lambda sets and other cohesive subsets. *Social Networks* 12 (4): 337–357.

Caddy, J.F. 2014. Why do assessments of demersal stocks largely ignore habitat? *ICES Journal of Marine Science* 71 (8): 2114–2126.

Caddy, J.F., and C. Stamatopoulos. 1990. Mapping growth and mortality rates of crevice-dwelling organisms onto a perforated surface: the relevance of 'cover' to the carrying capacity of natural and artificial habitats. *Estuarine, Coastal and Shelf Science* 31 (1): 87–106.

Coll, M., S. Libralato, S. Tudela, I. Palomera, and F. Pranovi. 2008. Ecosystem overfishing in the ocean. *PLoS One* 3 (12): e3881.

Collins, M., R. Knutti, J. Arblaster, J.L. Dufresne, T. Fichefet, P. Friedlingstein, X. Gao, W.J. Gutowski Jr., T. Johns, G. Krinner, M. Shongwe, C. Tebaldi, A.J. Weaver, M. Wehner, et al. 2013. Long-term climate change: projections, commitments and irreversibility. In *Climate change 2013-the physical science basis: contribution of Working Group I to the fifth assessment report of the Intergovernmental Panel on Climate Change*, 1029–1136. Cambridge, UK: Cambridge University Press.

Cury, P.M., and V. Christensen. 2005. Quantitative ecosystem indicators for fisheries management. *ICES Journal of Marine Science* 62 (3): 307–310.

Cury, P.M., L.J. Shannon, J.P. Roux, G.M. Daskalov, A. Jarre, C.L. Moloney, and D. Pauly. 2005. Trophodynamic indicators for an ecosystem approach to fisheries. *ICES Journal of Marine Science* 62 (3): 430–442.

Fedor, A., and V. Vasas. 2009. The robustness of keystone indices in food webs. *Journal of Theoretical Biology* 260 (3): 372–378.

Freeman, L.C. 1978. Centrality in social networks conceptual clarification. *Social Networks* 1 (3): 215–239.

Girvan, M., and M.E. Newman. 2002. Community structure in social and biological networks. *Proceedings of the National Academy of Sciences* 99 (12): 7821–7826.

Gulland, J.A., and S. Garcia. 1984. Observed patterns in multispecies fisheries. In *Exploitation of marine communities*, 155–190. Berlin, Germany: Springer.

Hairston, N.G., F.E. Smith, and L.B. Slobodkin. 1960. Community structure, population control, and competition. *The American Naturalist* 94 (879): 421–425.

Hemmingsen, A.M. 1960. Energy metabolism as related to body size and respiratory surfaces, and its evolution. *Reports of the Steno Memorial Hospital Copenhagen* 9: 1–110.

Huxel, G.R., and K. McCann. 1998. Food web stability: the influence of trophic flows across habitats. *The American Naturalist* 152 (3): 460–469.

Jennings, S. 2005. Indicators to support an ecosystem approach to fisheries. *Fish and Fisheries* 6 (3): 212–232.

Jennings, S., and N.V. Polunin. 1996. Impacts of fishing on tropical reef ecosystems. *Ambio* 25: 44–49.

Jordán, F. 2005. Topological key players in communities: the network perspective. *WIT Transactions on Ecology and the Environment* 81: 87–96.

Jordán, F., and I. Scheuring. 2004. Network ecology: topological constraints on ecosystem dynamics. *Physics of Life Reviews* 1 (3): 139–172.

Jordán, F., A. Takács-Sánta, and I. Molnár. 1999. A reliability theoretical quest for keystones. *Oikos* 86: 453–462.

Jordán, F., L. Wei-Chung, and A.J. Davis. 2006. Topological keystone species: measures of positional importance in foodwebs. *OIKOS* 112: 535–546.

Kleiber, M. 1932. Body size and metabolism. *Hilgardia* 6: 315–353.

Leontief, W.W. 1951. *The structure of the U.S. economy*. New York, NY: Oxford University Press.

Libralato, S., V. Christensen, and D. Pauly. 2006. A method for identifying keystone species in food web models. *Ecological Modelling* 195: 153–171.

MacArthur, R.H. 1955. Fluctuations of animal populations and a measure of community stability. *Ecology* 36: 533–536.

Nascimento, M.C., and L. Pitsoulis. 2013. Community detection by modularity maximization using GRASP with path relinking. *Computers & Operations Research* 40 (12): 3121–3131.

Newman, M.E., and M. Girvan. 2004. Finding and evaluating community structure in networks. *Physical Review E* 69 (2): 026113.

Oksanen, L. 1991. Trophic levels and trophic dynamics: a consensus emerging? *Trends in Ecology & Evolution* 6 (2): 58–60.

Oshima, M.C., and R.T. Leaf. 2018. Understanding the structure and resilience of trophic dynamics in the northern Gulf of Mexico using network analysis. *Bulletin of Marine Science* 94 (1): 21–46.

Pimm, S.L., and J.H. Lawton. 1980. Are food webs divided into compartments? *The Journal of Animal Ecology* 49: 879–898.

Polis, G.A., and D.R. Strong. 1996. Food web complexity and community dynamics. *The American Naturalist* 147 (5): 813–846.

Power, M.E. 1992. Top-down and bottom-up forces in food webs: do plants have primacy. *Ecology* 73 (3): 733–746.

Rezende, E.L., E.M. Albert, M.A. Fortuna, and J. Bascompte. 2009. Compartments in a marine food web associated with phylogeny, body mass, and habitat structure. *Ecology Letters* 12 (8): 779–788.

Saavedra, S., D.B. Stouffer, B. Uzzi, and J. Bascompte. 2011. Strong contributors to network persistence are the most vulnerable to extinction. *Nature* 478 (7368): 233–235.

Salcido-Guevara, L.A., and F. Arreguín-Sánchez. 2014. Effects of biomass changes in the supply–demand balance of energy in aquatic food webs. *Ecological Modelling* 276: 64–79.

Schaefer, M.B. 1954. Some aspects of the dynamics of populations important to the management of commercial marine fisheries. *Bulletin of Inter-American Tropical Tuna Commission* 1: 27–56.

Seidman, S.B., and B.L. Foster. 1978. A graph-theoretic generalization of the clique concept. *Journal of Mathematical Sociology* 6 (1): 139–154.

Shannon, C.E., and W. Weaver. 1949. *The mathematical theory of communication*. Urbana, IL: University of Illinois Press.

Stouffer, D.B., and J. Bascompte. 2011. Compartmentalization increases food-web persistence. *Proceedings of the National Academy of Sciences* 108 (9): 3648–3652.

Ulanowicz, R. 1986. *Growth and development: ecosystems phenomenology*. New York, NY: Springer, 203 p.

———. 2009. The dual nature of ecosystem dynamics. *Ecological Modelling* 220: 1886–1892.

Ulanowicz, R., S.J. Goerner, B. Lietaer, and R. Gomez. 2009. Quantifying sustainability: resilience, efficiency and the return of information theory. *Ecological Complexity* 6: 27–36.

Ulanowicz, R.E., R.D. Holt, and M. Barfield. 2014. Limits on ecosystem trophic complexity: insights from ecological network analysis. *Ecology Letters* 17 (2): 127–136.

Wassermann, S., and K. Faust. 1994. *Social network analysis*. Cambridge, UK: Cambridge University Press.

West, G.B., and J.H. Brown. 2005. The origin of allometric scaling laws in biology from genomes to ecosystems: towards a quantitative unifying theory of biological structure and organization. *Journal of Experimental Biology* 208 (9): 1575–1592.

West, G.B., J.H. Brown, and B.J., Enquist. 1997. A general model for the origin of allometric scaling laws in biology. *Science* 276: 122–126.

Zorach, A.C., and R.E. Ulanowicz. 2003. Quantifying the complexity of flow networks: how many roles are there? *Complexity* 8 (3): 68–76.

Chapter 4
Fisheries Within the Ecosystem Context: Production, Surplus Production, and Balanced Production

Production and Surplus Production

Within the context of the use of living resources, production refers to the biomass gained per unit of time by living organisms. In terms of use and the dynamics of natural systems, biomass use is based on the productive capacity of a population. Figure 4.1 illustrates the biomass production process. From top to bottom, the first graph illustrates (a) individual increases in weight with age (w_t) and (b) changes (gain) in weight per age (Δw_t). Figure (c) shows the typical decrease of a cohort, representing the number of individuals N_t of the population at age t. It follows that the biomass of the population is the product of the weight at each age, w_t, and the corresponding number of individuals, N_t; and the total production (P) would be the product of the increments of weight and the number of individuals, both with age, then

$$P = \sum_{t=1}^{n} N_t \cdot \Delta w_t$$

If there is an age at entry into the fishery, illustrated in Fig. 4.1 by the dotted line, the area under the production curve limited by the age at entry into the fishery will be the production capacity of the population that can be used. If the population is assumed to be undisturbed, and in equilibrium, with a stable carrying capacity, K, the production capacity of the population is only expected to replace losses due to natural causes. This state is represented as the asymptotic phase of the theoretical population growth curve in Fig. 4.2. Once the stock is exploited, the first effect is a reduced life expectancy of individuals, and the population must produce more biomass than what was necessary in the previous condition without disturbance in order to replenish the excess losses. Suppose that the exploitation phase begins at a relatively low harvest rate (HR), represented by state a of the population in Fig. 4.2. Maintaining an equilibrium scenario, the population must replenish an additional

F. Arreguín-Sánchez, *Holistic Approach to Ecosystem-Based Fisheries Management*, https://doi.org/10.1007/978-3-030-96847-2_4

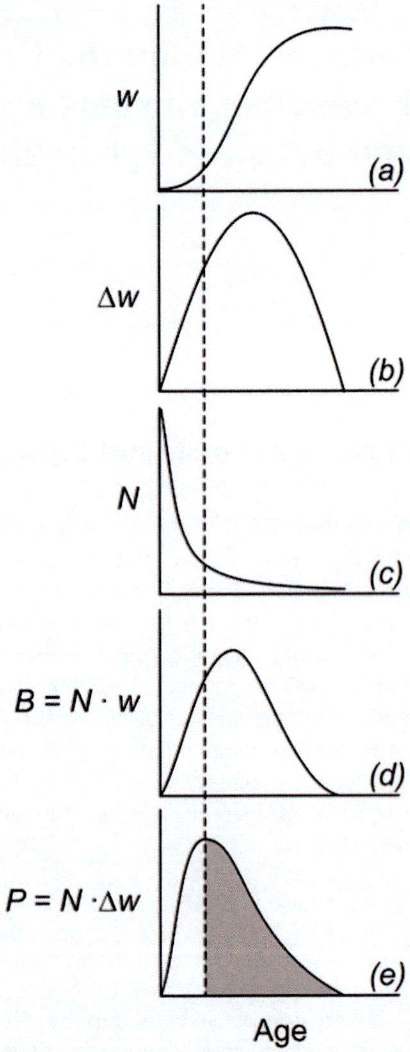

Fig. 4.1 Simplified graphical representation of the secondary production process for an individual stock, (*a*) represents individual growth in weight, w; (*b*) represents increments of biomass with age based on (*a*), Δw; (*c*) survival curve showing the change in numbers with age, N; (*d*) is the biomass with age for the entire stock, and is the product of numbers-at-age and individual weight-at-age, $B = N \cdot w$; (*e*) is the biomass production with age, $P = N \cdot \Delta w$. The dashed line represents the age-at-first capture in the fishery, and gray area represents the secondary production of ages vulnerable to fishing, which, in turn, represents the biomass available for fishing

amount of biomass, in this case equivalent to the area under the curve of that population's state *a* in the period of 1 year. If exploitation increases to state *b*, as shown in Fig. 4.2, the annual surplus production, as a response to the higher HR, will be greater. If the HR remains stable, the population will only be able to replenish, annually, the biomass represented by the surplus area according to the harvest state in which it is found; that is, assuming that the population is in state *b*, as shown in Fig. 4.2, it would take 6–7 years without harvesting for the population to recover to the carrying capacity *K*. This is the basic conceptual principle of exploited population dynamics models (i.e., Gulland 1969, 1983; Pauly 1983).

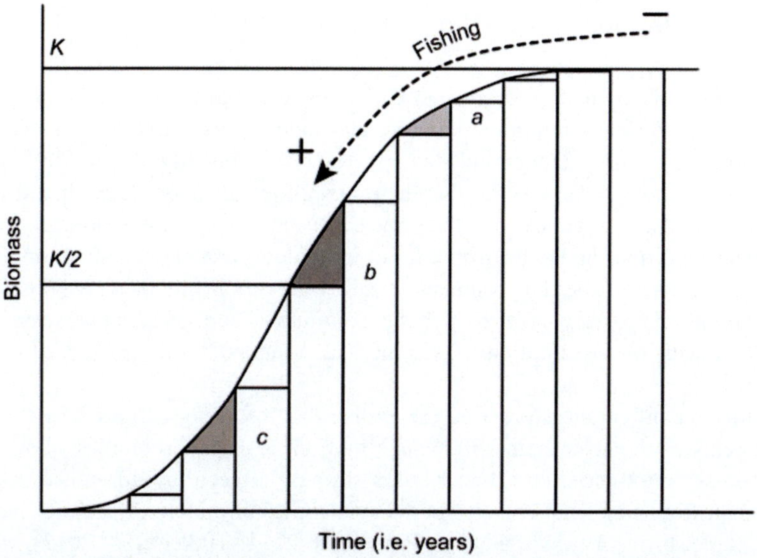

Fig. 4.2 Simplified graph that represents the production capacity of a population when exploited, corresponding to the potential annual yield or annual available biomass. *K* represents the load capacity of the system being the maximum size of the virgin population; Fishing reduces the size of the population and the life expectancy of fishes. The response of the population to biomass extraction is to increase production to restore annual losses, according to the amount of biomass extracted (triangles under the curve). In the figure, each vertical bar corresponds to an annual period, and the size of the triangle is the excess biomass produced by the population according to the impact of fishing. If fishing reduces the population to the level indicated by "a," the amount of biomass that the population is capable of recovering in 1 year will correspond to the gray rectangle. If the population is reduced until the *K*/2 level and maintained, the population will be in its state of maximum excess production capacity, indicated by the dark gray triangle to the left of the letter "b," and it is equivalent to the maximum sustainable yield, always that *K* and fishing intensity remains stable and unchanged. At this level, the exploited population is capable of replacing the biomass losses of the year represented by the dark gray triangle. If this level is exceeded, for example, the gray triangle to the left of the letter "c," the production capacity will no longer be enough to replace the losses because of fishing in 1 year, this being the condition of overfishing

The previous concepts, despite the assumptions of equilibrium and a stable carrying capacity (which also means stable ecosystems), were very useful for managing fishery resources, especially when the previous assumptions could be deemed reasonable. Under these conditions, fishing induced significant variation in population abundance trends. Consider again Fig. 2.1, which shows that these conditions were reasonably stable, with some variability, from 1930 to 1980. With the regime shift at early 1980s, a cooling period ends and a warming trend begins, and with this, the populations respond by adapting. Now, climate is also inducing changes in population trends, in addition to fishing.

Feeding Behavior and Sustainable Production

The response of populations to change varies according to their life histories, such that adapting to new conditions is not similar among them. If biomasses change continuously, so do trophic relationships and, subsequently, energy flows, as described in Chap. 2. This means that the probabilities of energy flows in a trophic network will also change, causing the same ecosystem to be reconfigured according to climate conditions and to the responses of individuals and of the population. If the inducing effect of the environment exhibits a certain trend over time, the reconfiguration process, in turn, will be continuous and, in a strict sense, not consistent with the assumption of stability, as assumed by the models used for decades.

In this reconfiguration process, the probability of energy flows between the components of the ecosystem will change with the changes in biomass; however, this process is not necessarily linear but is strongly affected by differences in diet preferences resulting from changes in the availability of preferred foods. Consider, for example, Fig. 4.3 (modified from Pires et al. 2011; Tinker et al. 2012), which shows three possible strategies of species in response to the scarcity of their preferred food.

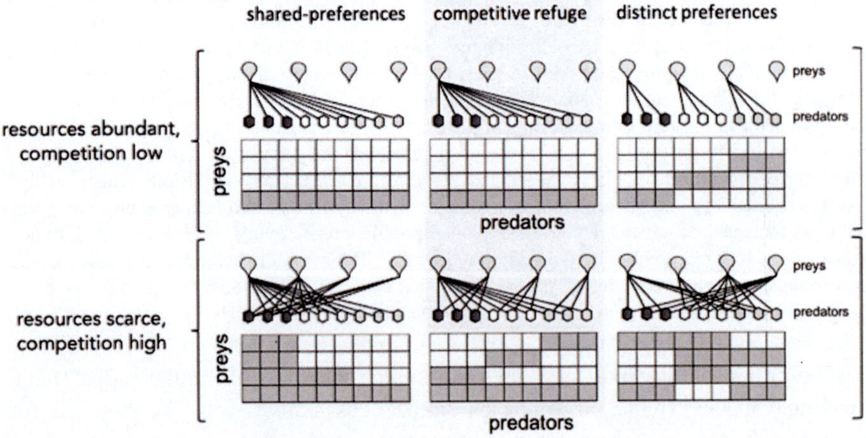

Fig. 4.3 Interspecific variation of the diet according to three theoretical models when resources are abundant (above) and when resources are scarce (below). The models represent conditions of shared preferences (left), competitive refuge (center), and distinct preferences (right). In the diagram, the circles represent prey and the polyhedra the predators; the lines define the trophic interactions. In the matrices, the lines are the prey (corresponding to the circles) and the columns to the predators (corresponding to the polyhedra), and the gray areas indicate the links. The models show different possible arrangements in the diets according to the availability of prey and the preferences and/or adaptability of the predators. These conceptual models help to explain the reconfiguration of ecosystems in the face of changes in climate patterns as a consequence of the adaptation of populations to the new conditions. This reconfiguration modifying ecosystem attributes related to its organization and functioning. (Adapted from Pires et al. 2011; Tinker et al. 2012)

The first case refers to the *competitive refuge* strategy, where species share a preferred resource but choose different alternative resources as the preferred food decreases. This process reduces diet overlap and increases niche breadth. The second case refers to the strategy where species show *distinct preferences* for food when their preferred food is scarce and may add a new type of resource to their diets. This behavior increases diet overlap. The third strategy, *shared preferences,* occurs when preferred food is scarce, species have similar preferences, but the rates at which they accept the second-choice resource vary, resulting in one species' diet being nested within that of another.

Let us now consider various species in a trophic network in which the probability of flow for a predator–prey relationship, $T_{ij}/T_{\circ\circ}$, changes according to any of the strategies of food exploitation described above where species adopt different strategies, at different times and under different circumstances, responding according to their vulnerability to a contingency, which in turn depends on their particular life history. If we link the above, i.e., the response capacity, which is different for each population in terms of the dynamics of the production of the populations illustrated in Figs. 4.1 and 4.2, it is clear that in a continuously changing environment with a directional trend, the possible persistence of the same stable ecosystem structure is unlikely, such that ecosystems will also continuously and gradually reconfigure.

The notion of sustainability in the use of fish resources and, therefore, of the ecosystems is often confused with the concept of stability. The relevant conceptual difference is that a stable system can be assumed to be sustainable by keeping adequate harvest rates reasonably constant. Under these conditions, the probabilities of energy flow in the food web also remain reasonably stable. On the other hand, in a process of continuous ecosystem reconfiguration, sustainability can be maintained by adapting the harvest rates to the production capacity of the species, which will depend on their particular trophic role in the ecosystem. This condition could be quantified from the changes in the probabilities of energy flow among the components of the ecosystem (trophic relationships), resulting in a balance between the harvest and production capacity of the species within the ecosystem, which will be different for each species. In this context, the system certainly does not remain stable, but stability in a system whose trend changes occur due to natural reasons is not a required condition nor is it a sustainable management goal.

Ecosystem Balanced Surplus Production

Garcia et al. (2012), recognizing the changes induced by biomass loss on the organization and function of ecosystems, suggest the concept of balanced catch. This concept is based on the general idea of proportionally taking biomass from different trophic levels without exceeding the production capacity of each level. Figure 4.4 illustrates this concept by representing the ecosystem as a trophic pyramid, where the pyramidal structure is unchanged, except for its total volume. Suppose that each level represents a certain amount of biomass; then, the idea would

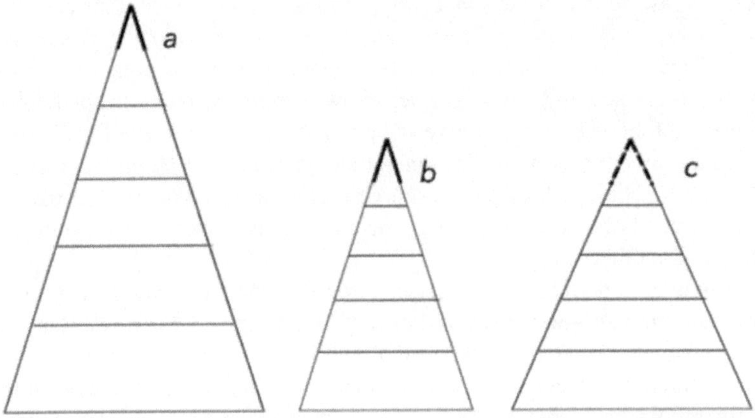

Fig. 4.4 Simplified graph representing ecosystem's states with trophic pyramids. Assuming that pyramid (a) represents a non-impacted ecosystem, the theory suggest that the best exploitation strategy could be such that no change its structure, function and organization, if this happens the angle in the top of the pyramids "a" and "b" should be equal. In contrast, if top predators are overexploited less amount of biomass will be transferred to upper trophic levels, making the base of the pyramid wider, and high trophic levels will be reduced and even could disappear (i.e., when fishing is highly intense). If such condition occurs, the angle on the top of the trophic pyramid will be larger than pyramid "a" representing the no-exploited trophic pyramid. This condition will affect structure, function, and organization of the ecosystem

be to use biomass in the same proportion from each trophic level to maintain its structure and functioning. This would imply a balanced harvest between trophic levels; of course, sustainability would be the level of use that allows the recovery of lost biomass, the ecosystem and each species in each level, per unit of time. Among the main drawbacks of this concept is that it would involve the harvesting of species that are not typically exploited (e.g., zooplankton, meiofauna) and that many species occupy different trophic levels throughout their ontogeny, such that maintaining a certain harvest rate at a low trophic level (i.e., recruits) could cause the collapse of the same species commercially harvested in a higher trophic level causing a low survival levels for adults; example known as growth overfishing.

The sustainable use of the production capacity of an ecosystem can be reasonably estimated from the population approach as long as the conditions of stability assumed in the models are acceptable; however, in the present time, this assumption can no longer be considered reasonable in general terms. To some degree, the concept of a balanced catch described above assumes an underlying stability because maintaining a stable trophic pyramid structure implies, in turn, the need to also maintain a stable probability of flows between trophic levels; at least, significant scientific input should be used to adjust the harvest rates of the different trophic levels year after year to maintain a stable pyramidal structure. In this case, the central precept is to maintain a pyramidal structure based on the principle that it is required to maintain the stable structure of the ecosystem, in this case the structure of the trophic pyramid, for a sustainable use of living resources; and of course, a

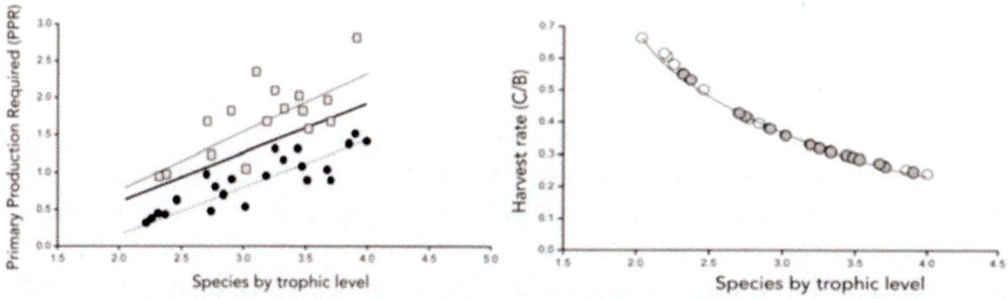

Fig. 4.5 For the figure on left, primary production required to sustain consumption, PPRQ, in the ecosystem (black dots) and to sustain fisheries, $PPRC_a$ (gray squares upper line) are represented. The theoretical consideration is that the fishery must use the resources in the ecosystem in the same way than nature, that is, as if it were one more predator, $PPRC_b$ Then the result is the black line (in the middle). Note that the slope of the PPRQ and $PPRC_b$ is the same separated by the surplus production resulted of the exploitation of each stock represented by their trophic level. Also note that the slope of $PPRC_a$ is higher. This means that top predators are being exploited at levels higher than defined by the natural balance expressed by the slope of line PPRQ. On the bottom, the harvest rates needed to maintain the balance between production capacity and extraction of biomass along the trophic levels are given. Note, as expected, that harvest rates for low trophic level are higher than for top predators. Example corresponds to the central Gulf of California

consequent question would be, what happens then with the reconfiguration of ecosystems due to changes in climate patterns? or are they little or nothing significant? Arreguín-Sánchez et al. (2022) suggest an alternative approach, as the balanced use of the production capacity of ecosystems, instead balanced catch. For any ecosystem model, based on a food web, it is possible to calculate the primary production required to support consumption by predators, PPRQ (Pauly and Christensen 1995). This estimation requires knowledge of the transfer efficiency and of the consumption magnitudes of the predator–prey ratios. In the same way, when some resource is used, the primary production required to support catches, PPRC, can be estimated. The central idea of the concept proposed by Arreguín-Sánchez et al. (2022) is that fishing must, at most, make use of (consume) the energy (biomass) of the food resources (prey species), in the same way as natural predators does and, in both cases, according to the production capacity of the populations, maintain ecosystem organization and functioning. Figure 4.5 illustrates the concept showing the relationship between PPRQ and PPRC for the different functional groups of the ecosystem, expressing species by their trophic level. The concept assumes that the pattern of resource utilization (i.e., fishing), expressed by the slope of the PPRC, should be similar to the PPRQ, which suggests that fishing uses the resources of the ecosystem in the same way as occurs in the natural processes of the ecosystem; that is, predation. However, the fishing limit would be defined by the ecosystem's maximum production capacity. To approximate this condition and given that the system is based on primary production, using the concepts of Fig. 4.2, the maximum possible production level of the lowest trophic levels would be defined by a maximum loss of 50% of its biomass. Thus, for example, the theoretical HR could be estimated for those lower trophic levels as:

$$\mathrm{HR} = C/B = 0.5 \quad \text{or} \quad \mathrm{HR} = \left(\frac{F}{(M+F)}\right)\left(1 - \exp^{-(M+F)}\right) = 0.5$$

and from this point, assume the same slope of PPRQ for $\mathrm{PPRC}_{\max}$. As shown in Fig. 4.5a, by proceeding in this way, the difference between the actual situation and the one in which the production capacity of the resources is maximized without altering ecosystem functioning can be identified. In the same way, the effect of overfishing, given by the difference in the value of the slope, can be identified, and for each particular case, it can be estimated as the difference $\mathrm{PPRC}_{i,\ \mathrm{actual}} - \mathrm{PPRC}_{i,\ \max}$. In addition, with the information on the $\mathrm{PPRC}_{\max}$, the optimal HR for each component of the ecosystem can be estimated, with particular interest for the exploited groups (Fig. 4.5b). In this approach, the HR corresponds to the limit biological reference point, $\mathrm{BRP}_{\mathrm{Lim}}$, for each species based on its production capacity, considering their trophic role in the ecosystem.

A concept of interest that emerges from the previous definitions, by way of analogy regarding the concept applied to exploited populations, is the surplus production, but in this case, for the whole ecosystem. Using Fig. 4.5a, for any species i, expressed by its trophic level, we can define its position on the line corresponding to PPRQ, and, consequently, its position on the line that represents the PPRC. In this context, the difference between these two points will be equivalent to the surplus production of species i in the ecosystem, defined as $\mathrm{PPRC}_i - \mathrm{PPRQ}_i$, and the ecosystem surplus production, ESP, will be:

$$\mathrm{ESP} = \sum_{i=1}^{n} \mathrm{PPRC}_i - \mathrm{PPRQ}_i \tag{4.1}$$

that is, the biomass equivalent to the area between the two straight lines that represent PPRC and PPRQ, where i represents the species and n represents the total number of species.

Overfishing Effects on Ecosystem Organization

On the other hand, the effect of overfishing a resource in the ecosystem can be defined through the changes in the probabilities of flow in the ecosystem at a level where the consumption of target species can no longer replace the loss of biomass. As mentioned in Chap. 3, when biomass is extracted from a population, the probabilities of energy flows are affected; in principle, these probabilities, represented by consumption, increase, in relative terms, as a consequence of the increase in the production of the exploited population (Fig. 4.2a, b). However, if the state of maximum production capacity is exceeded, the production rate decreases in such a way that consumption also decreases (Fig. 4.2c). This effect propagates in the food web directly and indirectly.

Following Ulanowicz (1986), *A/C* is a measure of the inherent organization of an ecosystem, such that:

$$A/C = \frac{\sum_{i,j} T_{i,j} \log \left(\frac{T_{i,j} T_{\circ,\circ}}{T_{i,\circ} T_{\circ,j}} \right)}{\sum_{i,j} T_{i,j} \log \left(\frac{T_{i,j}}{T_{\circ,\circ}} \right)} = \sum_{i,j} \left(\frac{[T_{\circ,\circ}]^2}{T_{i,\circ} T_{\circ,j}} \right) \tag{4.2}$$

and if *A/C* is a measure of the relative organization of an ecosystem, then $\mathcal{E} = 1 - (A/C)$ will be a measure of the entropy, $\mathcal{E}$, in the ecosystem. In this context, the entropy generated by biomass extraction is interpreted as degradation, in respect to the unperturbed system. If exploitation occurs within the production capacity of the populations and of the ecosystem, a certain degree of entropy is caused, but the populations, given their response capacity (according to the concept of Fig. 4.2), may eventually replenish losses or maintain a certain level of surplus production to counter the effects of fishing. On the other hand, if the level of entropy caused by extracting a certain amount of biomass exceeds the production capacity, the ecosystem will tend to degrade. This situation is exemplified in Fig. 4.6, which shows the trend of the change in $\mathcal{E}$ as the degree of species' biomass extraction increases. In these trends, the entropy of the ecosystem increases when the HR reaches a certain degree and then decreases, reaching an inflection point where the trend changes and eventually tends to stabilize. In this context, the inflection point is interpreted as the threshold of the entropy gained by the ecosystem as a result of biomass extraction, beyond which the ecosystem no longer returns to its initial state. One aspect of great interest is that the inflection points are generally within the same magnitude interval of the harvest rate, represented by the gray area in Fig. 4.6 at an HR of approximately 0.40.

According to Eq. (4.2), when extracting biomass from a population x and considering the concepts related to Fig. 4.3, it is likely that the relative consumption of the prey species that is being fishing target, $T_{x,\circ}$, tends to decrease as the biomass retained by fishing increases. In this same context, for the same target population, in its role as predator, the probabilities of relative consumption of prey y, $T_{\circ,y}$, will probably increase to support the surplus production. The effect of both the processes will propagate through the food web, and what could be assumed, in terms of the recovery of an exploited population within its productive capacity, is that $T_{\circ,y} > T_{x,\circ}$ (parts a and b of Fig. 4.2); if the state of the population is below the biomass necessary to generate a production capacity that allows it to replace the losses due to extraction, then $T_{\circ,y} < T_{x,\circ}$. This last condition will generate an imbalance in the exchange of flows that propagate through the trophic network, giving rise to an entropy gain in the system.

The carrying capacity for populations is defined by the limitations in their growth in biomass in the natural environment; these limitations may depend on various intra- and interpopulation factors, which in turn are mainly defined by trophic relationships and effects of the physical environment. In the context of an ecosystem, the carrying capacity corresponds to the total biomass of its component populations.

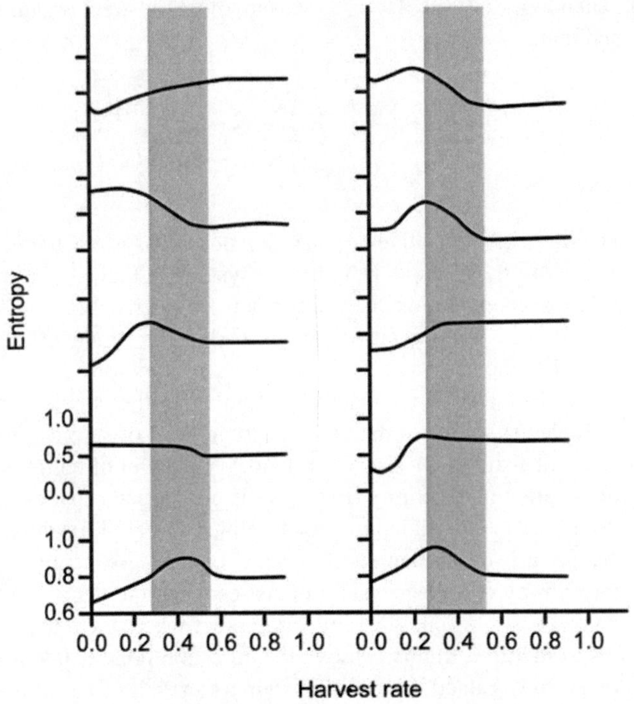

Fig. 4.6 Results of the simulation of effects of fishing on the organization and function of an ecosystem. Biomass extraction promotes changes in ecosystem entropy which in turn is an expression of the impact on ecosystem organization. Simulations are aimed to identify the limit harvest rate. During simulations, the harvest rate on a given species is gradually increasing and the change in entropy is measured. The inflection point of the change in entropy with harvest rate indicates the point of no return, identifying, as consequence, the limit reference level. The figure shows the trajectories of change in entropy for several species with the increase of harvest rates. Gray shadow area is to note that inflection points are located within a relatively small range of harvest rates. Simulations in picture come from different species or functional groups of the Campeche Bank ecosystem. From the left, from top to down, microcrustaceans, macroinvertebrates, porgies, anchovies, catfish. From the right, from top to down, mojarras, grunts, phytoplankton, zooplankton, shrimp

In both the cases, populations and ecosystem carrying capacities vary depending on the life histories of the populations, on the effects of environmental forcing, and on the trophic interactions, according to the ecosystem reconfiguration adopted when the biomass of the populations change and individuals adjust their feeding habits. However, if an ecosystem is exploited, as occurs with fisheries, the carrying capacity decreases when fishing is incorporated as an additional factor that limits growth, but the ecosystem is maintained in a condition of desired surplus production. In this case, the limits of the production capacity will depend on ecosystem.

References

Arreguín-Sánchez F., M.J. Zetina-Rejón, P. del Monte-Luna, M.O. Albáñez-Lucero, L.O. Duarte, D. Lercari, D. Calliari, R. Vogler. 2022. Reference harvest rates estimated from trophic level and production capacity of exploited stocks.

Garcia, S.M., J. Kolding, J. Rice, M.-J. Rochet, S. Zhou, T. Arimoto, J.E. Beyer, L. Borges, A. Bundy, D. Dunn, E.A. Fulton, M. Hall, M. Heino, R. Law, M. Makino, A.D. Rijnsdorp, F. Simard, and A.D.M. Smith. 2012. Reconsidering the consequences of selective fisheries. *Science* 335 (6072): 1045–1047.

Gulland J.A. 1969. Manual of methods for fish stock assessment. Part 1. Fish population analysis. *FAO manual on fisheries science*, No. 4, FAO, Rome, 154 pp.

Gulland, J.A. 1983. *Fish stock assessment: a manual of basic methods*. London, UK: Wiley.

Pauly, D. 1983. *Some simple methods for the assessment of tropical fish stocks (No. 234)*. Rome, Italy: Food & Agriculture Organization.

Pauly, D., and V. Christensen. 1995. Primary production required to sustain global fisheries. *Nature* 374 (6519): 255–257.

Pires, M.M., P.R. Guimarães Jr., M.S. Araújo, A.A. Giaretta, J.C.L. Costa, and S.F. Dos Reis. 2011. The nested assembly of individual-resource networks. *Journal of Animal Ecology* 80 (4): 896–903.

Tinker, M.T., P.R. Guimaraes Jr., M. Novak, F.M.D. Marquitti, J.L. Bodkin, M. Staedler, G. Bentall, and J.A. Estes. 2012. Structure and mechanism of diet specialisation: testing models of individual variation in resource use with sea otters. *Ecology Letters* 15 (5): 475–483.

Ulanowicz, R. 1986. *Growth and development: ecosystems phenomenology*. New York, NY: Springer, 203 p.

Chapter 5
Holistic Indicators: Theoretical Basis

Holistic Indicator and Ecosystem Performance

What is a holistic indicator? As mentioned previously, in the management of living marine resources, indicators are used as measures of the performance of a population and, depending on the management interest or objective, are usually expressed as reference points, either in the biological sense or economic sense. These reference points are also defined as target or limit reference points, depending on whether they characterize desirable or the threshold for a risky condition for an expected production. Ecosystems are characterized by a network structure, in which the maintenance of organization and function as a living system lies, in a global sense, on the balance between consumption and production capacities, which in turn, define attributes of a structural nature, such as the topology and cohesion, or on their function, such as resilience. In this sense, a holistic attribute refers to characteristics intrinsic to a network, in this case a trophic network, to its functioning, its organization and efficiency, and not to characteristics of individual populations. In this sense, a holistic indicator can also be defined in analogous terms as a reference point, either limit or target, depending on what is expected or desired from the use of the resources in an ecosystem and its production capacity and, in particular, its sustainability.

It should be kept in mind that ecosystems are dynamic systems whose structure and function change or vary continuously, and as a consequence, the performance indicators that characterize the network must represent this condition. The global attributes defined by Ulanowicz (1986), such as ascendency, development capacity, and overhead (see Chap. 2), constitute the theoretical basis for various attributes. Thus, for example, we find the concept of self-organize capacity (Ulanowicz 2009). According to Eq. (3.4), the self-organize capacity of an ecosystem represents a parabolic function, as shown in Fig. 5.1, which can be described as a generalized representation of the different types of ecosystems, such that towards the origin, where the self-organize capacity is minimal, the ecosystems show a low degree of

F. Arreguín-Sánchez, *Holistic Approach to Ecosystem-Based Fisheries Management*, https://doi.org/10.1007/978-3-030-96847-2_5

organization, but high overhead values, and exhibit therefore low efficiency. At the other extreme, ecosystems can also have little capacity for self-organization but be highly organized. This is where highly developed ecosystems are located, with low overhead since more of the energy flows pathways are limited (conditioned); they are rigidly linked systems and consequently are highly vulnerable to disturbances. In this context, the greatest capacity or power of self-organization corresponds to ecosystems with intermediate levels of organization, and in these ecosystems, there are more opportunities to contend with disturbances.

In an alternative context, this same diagram (Fig. 5.1) could allow identifying the changes in the capacity for self-organization of an ecosystem after being disturbed, where through changes in the degree of organization (gain of entropy), the impact on self-organize capacity could be identified, determining, as a consequence, the evolution of the ecosystem with respect to this performance indicator.

One aspect of interest regarding the capacity for self-organization of ecosystems is that this attribute is independent of changes in the structure of the ecosystem over time. Figure 5.2 shows an example of this based on simulation experiments for a marine-coastal ecosystem in the southern Gulf of Mexico. During the simulation of the trophic model of the ecosystem, environmental forcing related to climate change is incorporated, particularly on primary producers, for a period of 100 years, allowing the signal to propagate through the trophic network. For each year, the capacity for self-organization is estimated according to Eq. (3.4), showing, over time, the periods during which the capacity for self-organization, as a performance indicator, show the same value. Note how the structure of the ecosystem is different in each period, especially for high trophic level groups of consumers. As mentioned above, these results suggest that the indicator is independent of the structure and organization of the ecosystem, leading to its possible use as an indirect measure of

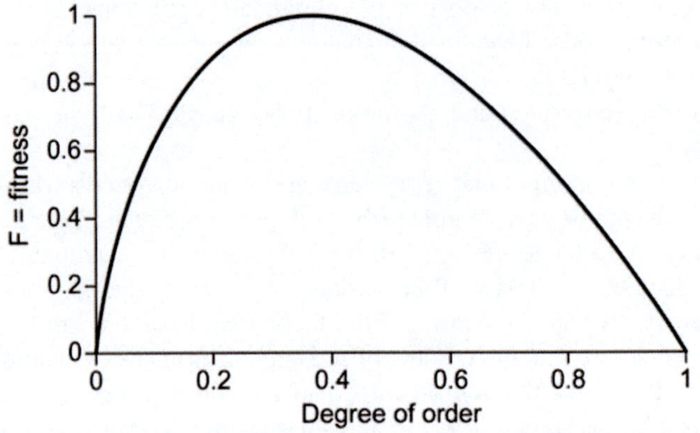

Fig. 5.1 The degrees of order and corresponding self-organization capacity given by the ecosystem fitness sensu Ulanowicz (2009). The higher self-organization capacity is present on intermediate degree of order. Less degree of order can be exemplified by low complex networks like upwelling ecosystems; while high degree of order is represented by high complex ecosystem like coral reefs

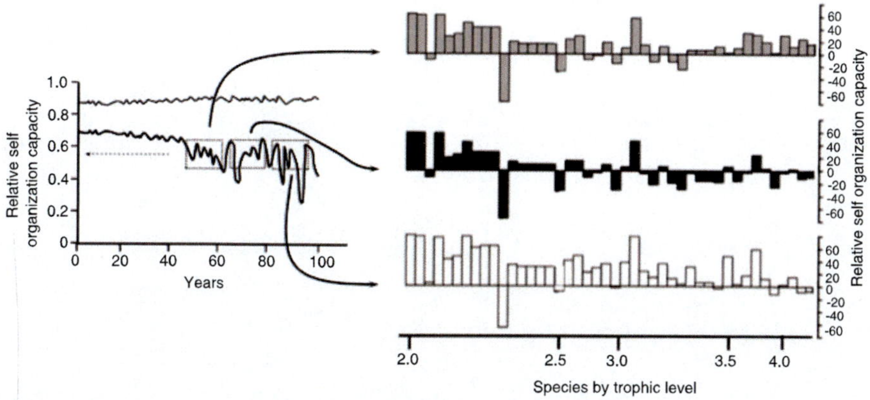

Fig. 5.2 Changes of the ecosystem self-organization capacity. Small picture on left represents changes in ecosystem fitness over time without climate effect (upper unchanged trend) and under effects of changes in climate patterns (lower trend). Note that when same values of fitness are present along the time, each condition did not correspond to the same ecosystem structure (right side panel), evidencing the effects more clearly towards high trophic levels. This suggests that ecosystems reconfigure themselves continuously under climate forcing occurs. (Adapted from Arreguín-Sánchez 2012)

resilience and/or sustainability in the face of processes such as climate change or other disturbances.

As seen in Fig. 5.2, the periods of time when the magnitude of the value of capacity for self-organization is similar correspond to different ecosystem configuration, where components varied in biomass. Considering the effects of climate change, as an example, and that ecosystems are continually reconfigured, this means simply different states of the ecosystem with similar organization attributes and, probably, resilience. In terms of the use of natural resources, this also suggests the need for an appropriate use of ecosystem holistic indices for assessing the adaptability of resource use strategies, in accordance to the ecosystem states along a continuous reconfiguration.

Ecosystem Window of Vitality

On the other hand, Zorach and Ulanowicz (2003) and Ulanowicz et al. (2009) propose the concept of a "window of vitality"when exploring the number of specialized roles or functions of species in an ecosystem, identifying this attribute as a measure of complexity, referring to weighted networks of which, trophic networks, are representative. The authors demonstrate that although the number of nodes (species) and flows vary widely, natural systems always fall within a small window of vitality with respect to the connectivity, *c*, and roles, *r*, of the components, where

$$c = \prod_{i,j} \left(\frac{T_{ij}^2}{T_{i\circ} T_{\circ j}} \right)^{-(1/2)\left(T_{ij}/T_{\circ\circ}\right)} \tag{5.1}$$

and

$$r = \prod_{i,j} \left(\frac{T_{ij} T_{\circ\circ}}{T_{i\circ} T_{\circ j}} \right)^{\left(T_{ij}/T_{\circ\circ}\right)} \tag{5.2a}$$

or

$$\log (r) = \sum_{i,j} \frac{T_{ij}}{T_{\circ\circ}} \log \frac{T_{ij} T_{\circ\circ}}{T_{i\circ} T_{\circ j}} = \mathrm{AMI} \tag{5.2b}$$

where AMI is the average mutual information, which is equivalent to the measure of ascendency weighted by total flows, AMI = $A/T_{\circ\circ}$ (defined in Eq. 3.8).

Zorach and Ulanowicz (2003) explore the properties of the window of vitality by relating the effective connectivity (link density), defined as the number of links that, on average, will flow into or out of a node in the trophic network; and the effective number of roles, defined as the quantity of different functions that occur in the network or number of transfers that a particle goes through in the system before leaving it.

Under the previous context, Zorach and Ulanowicz (2003) and Ulanowicz et al. (2009) define the limits of the vitality window as the limits of system sustainability. According to these authors, the heuristic interpretation of the limits of the vitality window (sustainability), along the ordinate axis, represents the link densities, where $c \geq 1$ indicates that the network is fully connected, and $c < 1$ indicates a disconnected network. The upper limit of this axis represents the maximum effective trophic level (Fig. 5.3a). With respect to the abscissa, values of $r \geq 2$ represent the limit of autotrophic/heterotrophic interactions in the system (Fiscus 2001). At the upper limit of the abscissa, the authors explain that the greatest possible value for n remains unexplained, although it is associated with the fact that in nature, values for trophic levels greater than 5 are rarely found. The result of this description is that the limits of the vitality window represent the limits of sustainability; and, given the dynamic nature of ecosystems, these limits will not necessarily exist at the point of maximum sustainability. Figure 5.3b represents the evolution of a northern Gulf of California ecosystem when forced by surface temperature anomalies as a consequence of climate change. The figure includes the historical series from 1900 to 2010 (Smith et al. 2008) and projections by the IPCC (2014) until 2100. The system shows oscillations along its trajectory approaching, at some point, the upper limits of both the axes. In this sense, the trajectory of the response to climate change represents a sustainable ecosystem that gradually evolves towards a greater prevalence of high trophic levels and greater connectivity, typical characteristics of a trend towards

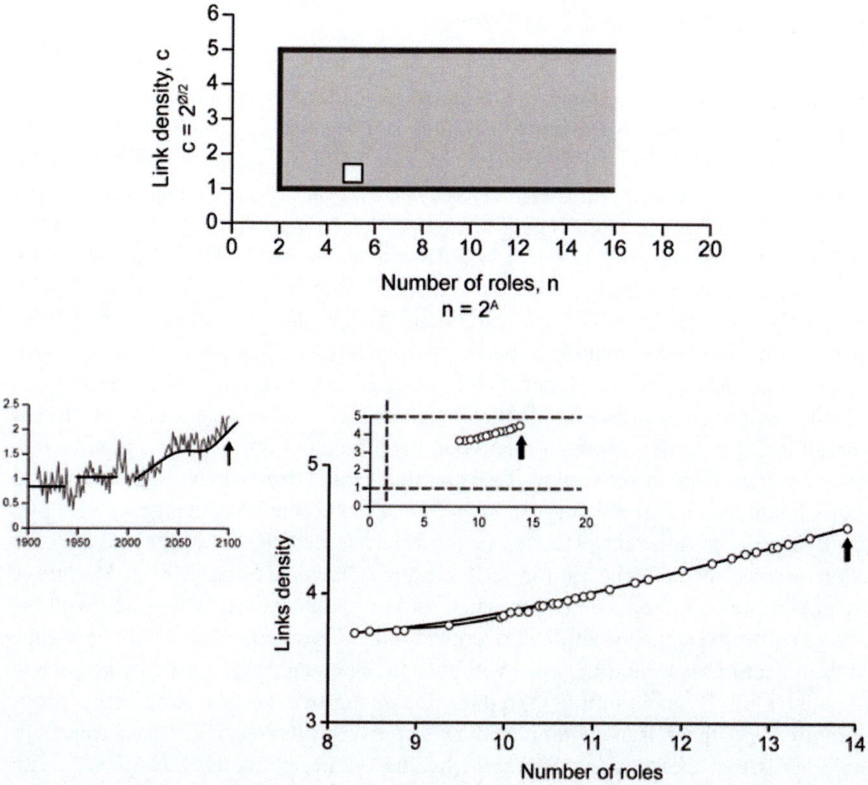

Fig. 5.3 Window of vitality sensu Ulanowicz et al. (2009) representing by the combination of link density and the number of roles for the components of the ecosystem. On the top, ∅ represents overhead and *A* Ascendency. Lower limit of link density, $c = 1$, represents a fully connected network, maximum-limit of *c* indicates ecosystems can be either strongly connected across a few links or weakly connected across many links. The lower limit of number of roles, $n > 2$, must be present for all ecosystems and represent the minimum level of interaction. The maximum level for *n* has not yet a proper interpretation. White square represents the maximum fitness for the ecosystem. Lower picture, on right, represents the trajectory of the ecosystem of the northern Gulf of California within the window of vitality when forcing primary production with the sea surface temperature, including both, historical, 1900 up to date, and projection until 2100 (left small picture)

greater complexity. The vitality window explored formally with observed data could be an excellent indicator of the sustainability of ecosystems and, in the same way, of the consequences of induced disturbances.

Ecosystem Health and Resilience

A concept that frequently appears in the scientific and managerial discourse is that of ecosystem health, which typically, although not clearly, is strongly linked to the concept of sustainability. The concept of sustainability seems to be reasonably defined in terms of energy flows and ecosystem organization through the concept of a vitality window. The concept of ecosystem health has been explored by several authors such as Ulanowicz (1980, 1992), Norton et al. (1992), Mageau et al. (1995, 1998), Rapport et al. (1998, 2009), Costanza and Mageau (1999), Costanza (2012). According to these authors, an ecosystem during its development (evolution) increases in flows, maintaining a balance between ascendency (representing vigor and organization of the ecosystem) and overhead (representing resilience and stability). Figure 5.4a represents this theoretical concept, showing a region of "fluxes variability" during development relative to the aforementioned balance that characterizes the health of an ecosystem. To better understand the concept, Fig. 5.4b shows a simulation, using a trophic model, of the effect of the change in mangrove cover on the dynamics of adjacent estuaries in the ecosystem of Bahía Magdalena on the western coast of the Baja California Peninsula (Chavez-Rosales 2006). Mangrove cover was gradually reduced (represented as percentages in the figure), and with the loss of mangroves, the system lost overhead; that is, resilience (horizontal tendency of the trajectory) was gradually lost without a major change in ascendancy (vigor and organization). By continuing to reduce the mangrove cover, ascendency drops abruptly, and there is a certain recovery of overhead. What this last trajectory suggests is that the ecosystem remains healthy while maintaining resilience, with

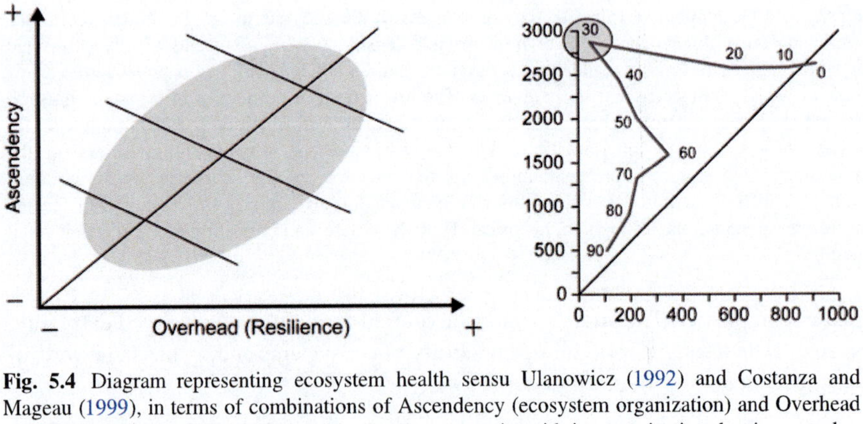

Fig. 5.4 Diagram representing ecosystem health sensu Ulanowicz (1992) and Costanza and Mageau (1999), in terms of combinations of Ascendency (ecosystem organization) and Overhead (resilience). The basic idea is that an ecosystem can persist with its organization despite perturbation, always that the limit of resilience was not exceeded. On the right, the effect of the loss of mangrove cover in the adjacent aquatic system for Magdalena Bay in the western coast of the Peninsula of Baja California. Numbers represent the percent of mangrove cover lost. At low levels of coverage loss, less than 30%, system can recover, even when resilience is diminishing. After such level, ecosystem cannot recover and organization is deteriorated with the increase in loss of mangrove coverage

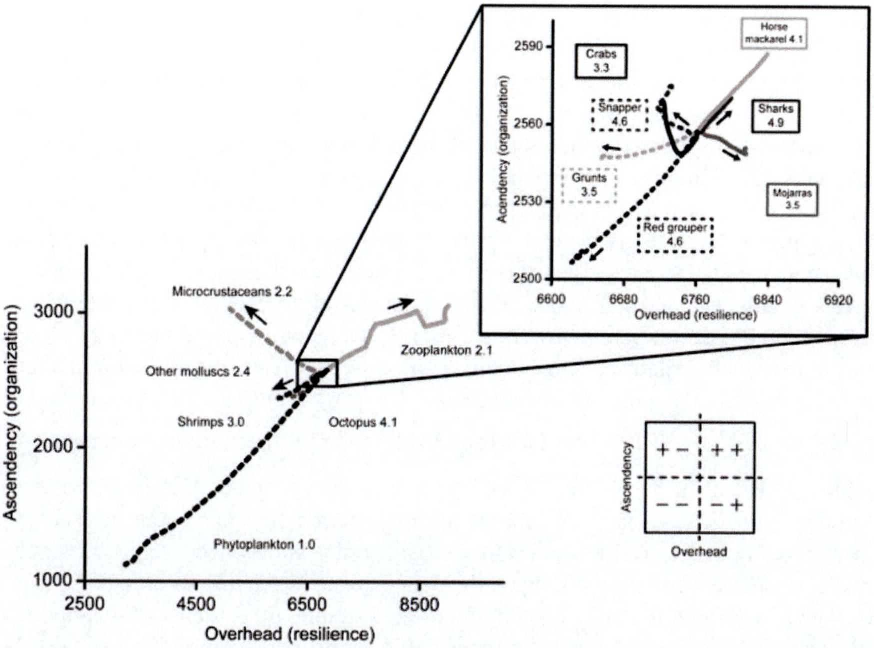

Fig. 5.5 Changes in ecosystem health due to changes in specific components for the continental shelf of Yucatan, Gulf of Mexico. Note that, in general, lower trophic levels cause larger effects than higher trophic levels indicating that ecosystem health is highly sustained by the former ones (see Fig. 5.4 for more details). The starting point represents the current ecosystem state in terms of ascendency (organization) and overhead (resilience). Each trajectory has three components: distance, representing resistance to perturbation; the slope representing organization-entropy balance; and trajectory direction (for explanation, see text)

the turning point being the limit reference point for the ecosystem, with a consequent loss of health after it. In other words, ecosystems can remain healthy despite disturbances, to the extent that they retain the resilience that allows them to recover their initial state.

In contrast, in Fig. 5.5, an example is shown for the continental shelf ecosystem of the Yucatan Peninsula, in the Gulf of Mexico, where the biomass of each component, individual and independent of each, gradually reduced from 0 to 90%, measuring in each case the ascendency and the overhead. In this context, the magnitude of the effects on the different components of the ecosystem according to their position and trophic role is highlighted. As observed in Fig. 5.4 (right), the horizontal trajectories in Fig. 5.5 suggest a loss of resilience, while the trajectories towards the origin represent high decomposition in the ecosystem because both ascendency and overhead are lost.

A concept closely linked to resilience is resistance, that is, the time it takes for an ecosystem to respond to a disturbance. In this context, taking as an example Fig. 5.4 (right), if the gradient represents the decrease in mangrove cover over time, it is likely that the time elapsed along the horizontal trajectory, until the point of inflection, could be directly associated with the ecosystem resistance. On the other hand, the response to the disturbances induced during the simulations (Fig. 5.5) illustrates three notable aspects. (1) The distance traveled from the starting point represents the magnitude of the change required by a group to affect the ecosystem; a greater distance implies a lower relative impact, characteristic of short-lived species and high recovery rates; in high trophic-level species, the distance is considerably less, typically corresponding to long-lived species with low recovery rates. This behavior is related to the resistance of the ecosystem, being greater when low trophic levels are affected, and vice versa; i.e., the resistance of the ecosystem is lower when the disturbance occurs at high trophic levels. (2) The second aspect refers to the slope that represents the relative change in the order of the ecosystem regarding the change in the overhead or potential for resilience (see Chap. 6), represented in terms of flows as $\beta_{s,i} = -\frac{\log\left(T_{ij}T_{\circ\circ}\right)}{\log\left(T_{ij}^2\right)}$. (3) The third aspect refers to the direction of the trajectory, which represents four possible states of the ecosystem responding to the disturbance. Starting from the base state, the ecosystem can gain order and gain overhead (energy in reserve, linked to resilience), or it can gain order but reduce overhead, which makes it a more vulnerable system. A third state occurs when it loses order but gains resilience; although this may be considered undesirable, the system can respond to the disturbance given the gain in resilience. The fourth state represents a scenario in which the ecosystem has been strongly disturbed and losing both order and overhead, i.e., a system in degradation state.

In Fig. 5.5, several aspects are of interest; the direction that the ecosystem takes when phytoplankton is affected is one of frank degradation (overhead and ascendency are reduced), although the distance relative to the magnitudes of change is the largest among the functional groups that make up the ecosystem. This is because phytoplankton is at the base of the trophic network. These same tendencies of frank degradation are observed in top predators, with the difference that the magnitudes of change and distance are clearly lower. This suggests that the ecosystem is highly vulnerable to impacts at higher trophic levels, which is explained, according to Arreguín-Sánchez and Ruiz-Barreiro (2014), in addition to the presence of long-living species, to the dominance of the top-down control faced of a limited ecosystem carrying capacity. In contrast, the opposite scenario occurs with zooplankton, where the system gains organization and resilience as biomass is reduced. This trajectory suggests that as zooplankton biomass loss increases, the limitation of flows to the rest of the system increasing resilience.

The four states of the ecosystem mentioned above can be identified by following the trajectory of the ecosystem represented in Fig. 5.5; the overhead gain, in general, coincides with the region of maximum organization, and the direction of $\beta_{s,\,i}$ suggests that the system tends to increase or decrease its level of organization. On the other hand, the decrease in overhead can reflect both the simplification of the ecosystem by losing order (less limitation of flow and less order) or by gaining order and complexity as a result of the increase in the limitation of flows (e.g., increase the

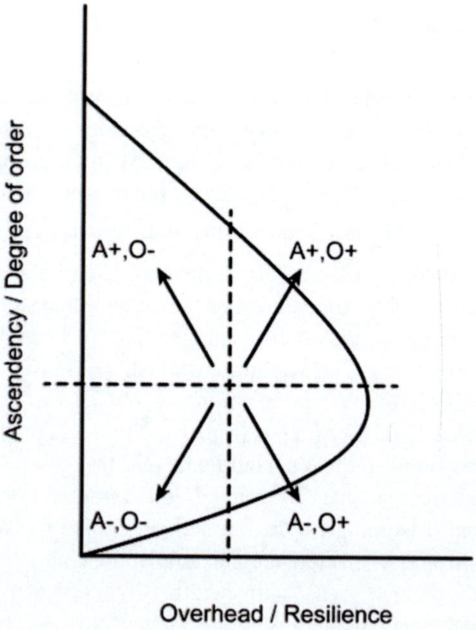

Fig. 5.6 Diagram explaining the combination of ascendency and overhead for different conditions of ecosystem health. Arrows pointing up represent gain in ecosystem organization, with increase or decrease in resilience (right and left, respectively). By the contrary, the loss of organization is represented by arrows with down direction, and also, right and left represent increase or decrease in resilience, respectively. Evidently, the worst ecosystem state is when both organization and resilience decreased; while the better condition of development could be, theoretically, represented by increase in both organization and resilience; however, in nature such state, if present, could be transitory, because a high developed ecosystem is characterized by using most of the available energy in organization leaving low available energy for overhead

number of speciation events). These conditions and states are conceptually shown in Fig. 5.6.

Of course, the above conclusions refer to the response of the ecosystem to impacts that affect the biomass of specific components of the system, one at a time, independently of each other. Under these conditions, response trajectories regarding changes in order and entropy allow the exploration and understanding of some particular aspects of ecosystem behavior and possible responses to disturbances. In nature, however, changes in and evolution of ecosystems occur differently, and populations vary simultaneously and in different ways according to the nature of the disturbances, the dynamics of interdependencies in the trophic network and the particular state of the ecosystem in which it is found, understanding, of course, that the responses in a complex system are not linear.

Ecosystem Vulnerability

On the other hand, if the trophic level of an ecosystem component, TL_i (as an index of the role of each component of the ecosystem), and distance, dist_i, of the trajectory traveled during the disturbance simulations (Fig. 5.5) (relationship between ascendency and overhead) are considered, weighted by the magnitude of the slope $\beta_{s,\, i}$, an ecosystem vulnerability indicator can be obtained, represented as $\nu_i = \beta_{s,i}\left(\frac{\mathrm{TL}_i}{\mathrm{dist}_i}\right)$, where ν_i refers to the relative vulnerability of the ecosystem relative to the reduction in biomass (disturbance) of component i. For the entire ecosystem, the relationship between ν_i vs. TL_i can be found, where the slope represents a global indicator of the vulnerability of the ecosystem. This condition is illustrated in Fig. 5.7 for two marine ecosystems in the Gulf of Mexico where, according to what was described above, the system in the southern Gulf of Mexico (right) is more vulnerable than the north continental shelf ecosystem of the Yucatan Peninsula (left).

One aspect of interest is that, in general, the removal of biomass from an ecosystem component (assuming for the moment an undisturbed ecosystem) generates an increase in entropy. In this sense, for example, the changes represented by the trajectories in the diagram of ecosystem health (Fig. 5.5) can be represented as changes in how orderly an ecosystem is, represented as $a = A/C$ (see Eq. 3.4); given that $C = A + \phi$, where C is the development capacity of the ecosystem, and ascendency, A, and overhead, ϕ, are the components of the diagram in Fig. 5.4. In this way, if A/C represents the relative order of the ecosystem, $1 - A/C$ will represent the relative entropy. Thus, Fig. 5.8 represents the same trajectories in Fig. 5.5 in

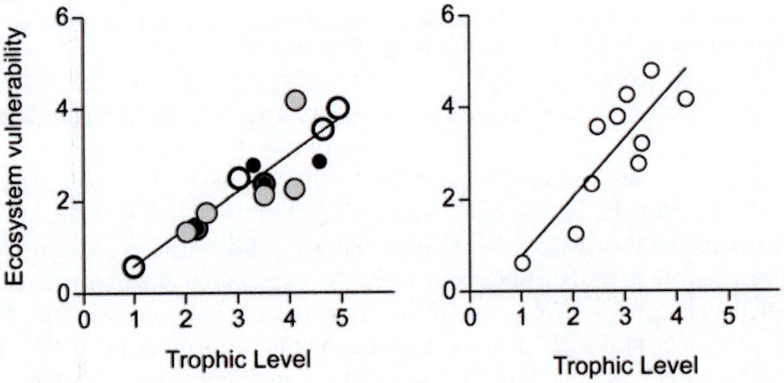

Fig. 5.7 Relative vulnerability of ecosystems of the North continental shelf of Yucatan (left) and southern Gulf of Mexico (Campeche Sound). Each point represents the relative vulnerability of the ecosystem $\mathcal{V}_{\mathrm{TL}}$ when a given group, represented by his trophic level, TL, is perturbed following Fig. 5.5. Each $\mathcal{V}_{\mathrm{r}} = \log\left[\beta_{s,i}\left(\frac{\mathrm{TL}}{R}\right)\right]$, where $\beta_{s,i} = -\frac{\log\left(T_{ij}T_{\circ\circ}\right)}{\log\left(T_{ij}^2\right)}$; and R the ecosystem resistance represented by the distance of the perturbation as explained in Figs. 5.5 and 5.8. The slope in each figure is an index of the ecosystem vulnerability, being $b = 0.788$ and $b = 0.689$, for Yucatan and Campeche ecosystems, respectively

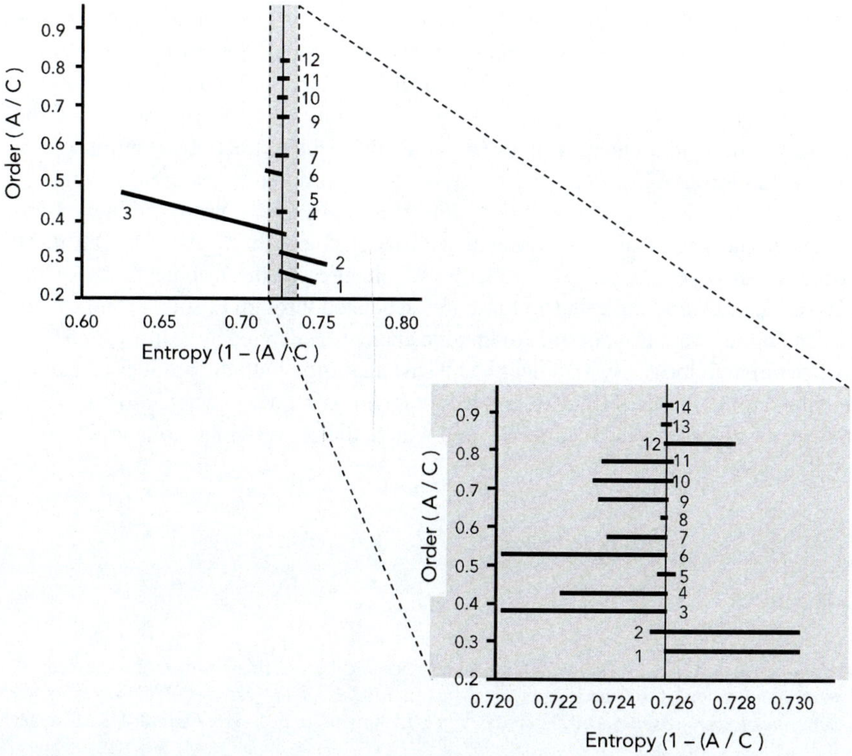

Fig. 5.8 Relationship between changes in order and entropy in the ecosystems of the North continental shelf of Yucatan (left) and southern Gulf of Mexico (Campeche Sound) when each functional group in the ecosystem is perturbed. The length of the lines represents ecosystem resistance, and direction, left-right, represents the loss or gain in entropy when a given group is perturbed. (Adapted from Arreguín-Sánchez and Ruiz-Barreiro 2014)

terms of changes in order and entropy. In this case, the numbers represent the components of the trophic network of the ecosystem, starting with the primary producers; and the distances represent the magnitudes of entropy gain or loss according to the effects of removing biomass from the ecosystem (in the conditions in which the simulation experiment was developed).

If this representation is compared with Fig. 5.5, there will be clear differences in the responses, given that they represent different attributes of ecosystem dynamics. Thus, for example, in Fig. 5.8, groups 1 and 2 represent phytoplankton and zooplankton, respectively. In this case, the effects of the disturbance caused by the gradual decrease in biomass generate a gain in entropy, that is, ecosystem degradation, and the distance corresponds, in this case, to the magnitude of the entropy generated; such that the entropy gained is not related to the resilience of the ecosystem, an attribute that is represented by overhead in the ecosystem health diagram.

In this context, the robustness attribute is represented by the relationship $R = T_{\circ\circ} \cdot F$ (see Eqs. 3.4 and 3.5), where $T_{\circ\circ}$ represents the total trophic flows of the ecosystem and F represents the capacity for self-organization, which in turn is a function of the ecosystem order, $a = A/C$. In this way, the changes in the robustness of the ecosystem due to the changes in flows induced by the extraction of biomass can be estimated through Eq. (3.5) and is inversely proportional to the entropy gain.

On the other hand, it is possible to link the robustness, entropy, vulnerability, resistance, and health of an ecosystem with its sustainability, given that all these attributes can be explained as a function of changes in the trophic flows of the ecosystem; and these attributes, in turn, are associated through Eqs. (5.1) and (5.2a). In an analogous way, it is possible to link the changes generated by entropy, and their relationship with the ecosystem health and sustainability, with the production capacity of the populations in the ecosystem where the maximum acceptable biomass to be retained out of the ecosystem; and, beyond which, the ecosystem no longer returns to its initial state.

References

Arreguín-Sánchez, F. 2012. The dynamics linking biological hierarchies, fish stocks and ecosystems: implications for fisheries management. In *Models of the ecological hierarchy: from molecules to the ecosphere*, ed. F. Jordán and S.E. Jørgensen, 501–516. Oxford, UK: Elsevier B.V.

Arreguín-Sánchez, F., and T.M. Ruiz-Barreiro. 2014. Approaching a functional measure of vulnerability in marine ecosystems. *Ecological Indicators* 45: 130–138.

Chavez-Rosales, S. 2006. *El papel de los manglares en la producción de las comunidades acuáticas de Bahía Magdalena*. B.C.S. Tesis Doctor en Ciencias, Instituto Politécnico Nacional, Centro Interdisciplinario ee Ciencias Marinas, La Paz, Baja California Sur, México, 143 p.

Costanza, R. 2012. Ecosystem health and ecological engineering. *Ecological Engineering* 45: 24–29.

Costanza, R., and M. Mageau. 1999. What is a healthy ecosystem? *Aquatic Ecology* 33 (1): 105–115.

Fiscus, D. 2001. The ecosystemic life hypothesis I: introduction and definitions. *Bulletin of the Ecological Society of America* 82 (4): 248–250.

IPCC. 2014. Climate change 2013: the physical science basis. In *Working Group I contribution of to the fifth assessment report of the Intergovernmental Panel on Climate Change*, ed. T.F. Stocker, D. Qin, G.K. Platner, M.M.B. Tignor, S.K. Allen, J. Boschung, A. Nauels, Y. Xia, V. Bex, and P.M. Midgley. Cambridge, UK: Cambridge University Press, 1537 p.

Mageau, M.T., R. Costanza, and R.E. Ulanowicz. 1995. The development and initial testing of a quantitative assessment of ecosystem health. *Ecosystem Health* 1 (4): 201–213.

——— 1998. Quantifying the trends expected in developing ecosystems. *Ecological Modelling* 112 (1): 1–22.

Norton, R.C.B.G., M. Faber, and D. Rapport. 1992. *Ecosystem health: new goals for environmental management*. Washington, DC: Island Press.

Rapport, D.J., R. Costanza, and A.J. McMichael. 1998. Assessing ecosystem health. *Trends in Ecology & Evolution* 13 (10): 397–402.

Rapport, D.J., C.L. Gaudet, R. Constanza, P.R. Epstein, and R. Levins, eds. 2009. *Ecosystem health: principles and practice*. New York, NY: Wiley.

Smith, T.M., R.W. Reynolds, T.C. Peterson, and J. Lawrimore. 2008. Improvements to NOAA's historical merged land-ocean surface temperature analysis (1880-2006). *Journal of Climate* 21 (10): 2283–2296.

Ulanowicz, R.E. 1980. An hypothesis on the development of natural communities. *Journal of Theoretical Biology* 85 (2): 223–245.

Ulanowicz, R. 1986. *Growth and development: Ecosystems phenomenology*, 203. New York, NY: Springer.

Ulanowicz, R.E. 1992. Ecosystem health and trophic flow networks. In *Ecosystem health: new goals for environmental management*, ed. R.C.B.G. Norton, M. Faber, and D. Rapport, 190–206. Washington, DC: Island Press.

———. 2009. The dual nature of ecosystem dynamics. *Ecological Modelling* 220 (16): 1886–1892.

Ulanowicz, R.E., S.J. Goerner, B. Lietaer, and R. Gomez. 2009. Quantifying sustainability: resilience, efficiency and the return of information theory. *Ecological Complexity* 6 (1): 27–36.

Zorach, A.C., and R.E. Ulanowicz. 2003. Quantifying the complexity of flow networks: how many roles are there? *Complexity* 8 (3): 68–76.

Chapter 6
Ecosystem Resilience

Panarchy, Socioecological Systems, and Practice of Fishing

In ecology, the resilience of a living natural system is, in the broadest sense, its ability to recover after having suffered a disturbance. While this concept can be widely accepted, in regard to resource management, in our case aquatic and, in particular, exploited ecosystems, it is important to recognize the dynamics of processes and the scales of complexity. Holling (1973) defined the resilience of an ecosystem as the magnitude of disturbance that the system can experience before transforming to an alternate state (referring to stability) with different structures and functioning.[1] Folke et al. (2004), Folke (2006) and Folke et al. (2010) suggest that resilience is the ability of an ecosystem to absorb a disturbance and reorganize itself while undergoing changes to maintain its function, structure, and identity. Of course, the ability to reorganize and renew an exploited ecosystem depends on the state in which the system is to be maintained but largely on the characteristics and nature of ecological hierarchies and their interactions. The recognition of the dynamics of these organizational scales and their interdependencies have been considered in the concept of panarchy (Gunderson and Holling 2002), which is key for the use and sustainability sought in the utilization of fishery resources.

Let us consider this concept of panarchy in greater detail. Figure 6.1 shows different levels or hierarchies of biological/ecological, and social complexity when considering humans as part of the natural system. Faced with a disturbance, say environmental, individuals of a population of fish of commercial interest respond by adjusting their metabolism, manifesting in aspects such as biomass gain, or their physiology, affecting, for example, gonad maturation or individual growth.

[1] Holling (1973) also mentions the concept of engineering resilience, referring to the rate at which a system restores itself to its initial state, but does not recommend the use of this concept because natural systems tend to have multiple unstable states. This concept of engineering resilience does not therefore seem very useful in terms of management.

F. Arreguín-Sánchez, *Holistic Approach to Ecosystem-Based Fisheries Management*, https://doi.org/10.1007/978-3-030-96847-2_6

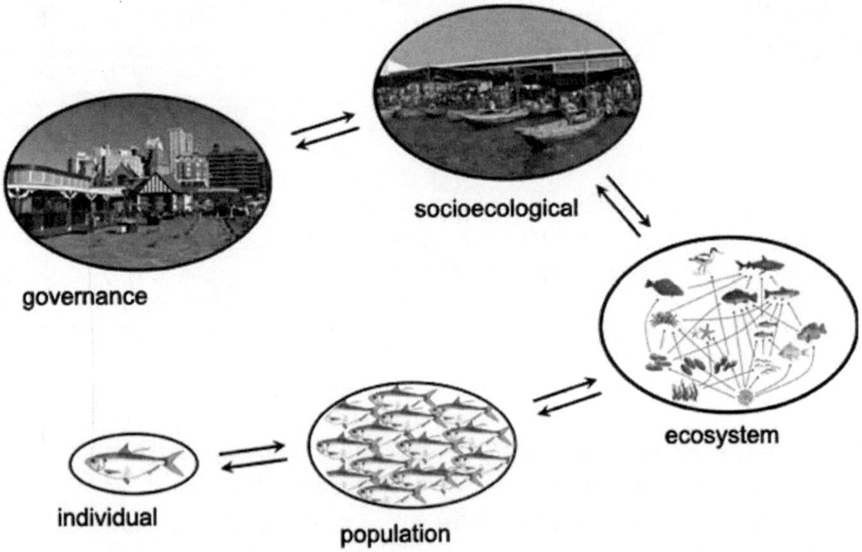

Fig. 6.1 Diagram representing the dynamics among biological and socioeconomical hierarchies related to fishing, and taken as base for management. An individual fish is a natural system whose performance is expressed by their metabolism in terms of robustness, gonadal maturity, among others. The population as a system represents the dynamics of a group of individuals whose performance is given by survival and related processes. Populations dynamically interact, each other in different ways given place to ecosystems, whose performance is measured through network attributes referred to its organization and function. The human communities in the coasts use natural resources of the ecosystem, extracting biomass and taking income and food. The sustainable use of these resources represents their livelihoods. The socio-ecological system performance can be measured by different indices like net rent, food security, reduction of poverty among others; and the management of all these systems is represented by the governance system which is constituted by entities of different nature that participate in the fishing sector, and whose performance can be measured through human development indices, efficiency of management measures, etc. The main concept, in addition to the dynamics within and between the systems is that they are interdependent on each other and that their relationships are not linear and, under events like climate change, the knowledge supporting management must be continuously provided

Individuals, according to their condition, adjust their metabolism, adapt and respond, either positively or negatively, according to the favorable or unfavorable effect of the disturbance. This response defines their resilience, and the metabolic efficiency of individuals can be defined by the gain in biomass or robustness and the performance of their physiology, attributes that, under a disturbance, are altered. However, individuals make up a population. The processes at population level are not the sum of individual processes; the population has its own dynamics and its own processes. In general, the efficiency of a population can be measured by its persistence, that is, its survival and maintenance. Processes such as reproductive success can be gauged through success in recruitment and maintenance of the population structure relative to survival; these are probably the key elements through which resilience acts at this level of complexity. In other words, the compensatory processes in a disturbed population (i.e., faster growth and greater survival at low population levels, sensu Cushing 1975, 1995) are a manifestation of their resilience. Of course, individual and population processes are linked. Each natural system, according to its complexity, has its own mechanisms and processes, in addition to the interdependence between them.

Aquatic ecosystems are considered complex entities where populations interact with each other, exchanging energy through predator–prey relationships and maintaining a dynamic balance between populations, which in turn, maintain dynamic stability in the ecosystem. In terms of efficiency, the ability of ecosystems to absorb the effects of disturbances and restore their organization and functions is what defines their resilience. Thus, for example, when a population is exploited, the ecosystem functions as a support; that is, the ecosystem, through trophic flows, provides the necessary energy for the population to incorporate it into its metabolism and be able to respond and restore losses according to their intrinsic life history characteristics. Naturally, a reduction in biomass would propagate throughout the ecosystem, directly and indirectly, generating a gain in entropy as a global response of the system; thus, the ability of an ecosystem to restore its organization is a measure of resilience. The stronger the disturbance is, the greater the entropy affecting the organization of the ecosystem, which is restored in accordance with the dynamics of the ecosystem. In this context, of course, the differential role that each population plays within the ecosystem comes into consideration, thus manifesting in this way the interdependence between the two levels of ecological hierarchy.

The next level of hierarchy is the socioeconomic system, which essentially reflects the interaction between the natural system and the human population that depends on the resources used. In this context, resilience refers to the capacity of the human sector to adapt and cope with changes in the availability of resources. Commercial fishing is essentially an extractive activity, and success is measured by the benefit or well-being of the dependent human population, in terms of either food availability or income. According to Herrera and Rodríguez (2016), in this context, resilience can refer to a system's dynamic capacity for learning and adaptability to self-organize without changing its essential functions in response to disturbances, allowing innovation, development, and evolution. In socioeconomic systems, the human population interacts with resources of an ecosystem, where harvesting strategies based on the availability of resources define the interaction between natural and human systems. The resilience of the socioeconomic system will depend largely on maintaining the dynamic balance of these interactions. When the adaptability of the human system is conceived as a form of resilience, two particular attributes are manifested: the existence of memory or learning and the provision of technology, with both favoring a greater capacity of adaptability that can be modified (typically to increase such capacity) to a certain degree responding to the same type and magnitude of impact. In this case, resilience also has an objective, generally linked to social welfare.

The next level of complexity or hierarchy (Fig. 6.1) is governance systems, which refer to the institutional arrangements that make resource management possible. In this context, the term "institutional" refers to various entities that participate in the organization and functioning of decision-making systems for the administration and control of fishing, e.g., nongovernmental, governmental, productive or marketing groups, academic-scientific entities, control and surveillance, and regulations, that is, how the administrative and management system, under specific objectives, is organized and functions to achieve the expected goals of resource use. In governance systems, planning and decision-making are geared towards allocating the benefits derived from the use of resources and, to the extent possible, being adaptable so that the disturbances, the socioeconomic system, the natural ecosystem and the populations can maintain a dynamic state of balance that ultimately ensures the sustainable use of fishery resources. As in the previous case, the resilience of governance systems refers, sensu Herrera and Rodríguez (2016), to the capacity of institutions and/or entities to adapt to and cope with changes in the availability of resources to achieve the goals defined regarding the allocation of benefits derived from the use of resources.

Figure 6.1 shows the complexity gradient associated with the hierarchies of interdependent systems, where for example, individuals as independent systems are a structural and functional part of the population, the next level of complexity; while populations are themselves embedded within the ecosystem at the next hierarchical level, and so on, with the socioeconomic and governance systems. In this way, in the dynamics of fishing, as an activity belonging to the primary production sector in the economy, resilience, as a global concept, acts in different ways within the hierarchy that corresponds to each system, as well as between hierarchies, since each system is embedded in the next hierarchical level of complexity. Similarly, each individual system shown in Fig. 6.1 assumes its own complexity due to its internal dynamics and a form of resilience as an intrinsic property but dynamically linked to the systems (hierarchical levels) with which it is interdependent.

Adaptative Cycle Sensu Holling (1986)

According to Holling (1986, 2001), in the processes of change in complex systems, four states can be characterized as a result of their internal dynamics and external influence, known as the adaptive cycle (Fig. 6.2). The importance of this concept lies in two aspects. First, theoretically, resilience is an intrinsic part of the process in each of the four phases. Second, in terms of resource management, in a variable environment where organisms respond to external stimuli, populations adapt and ecosystems are reconfigured, and therefore, management cannot be static; and this must be dynamic and susceptible to adaptability, which, to a large extent, is associated, at least regarding risk and probability of success, with the resilience of the exploited natural system.

Consider Fig. 6.2; and suppose a natural system, for example, a population; and that the adaptive cycle begins with the growth phase. Within this phase, the system

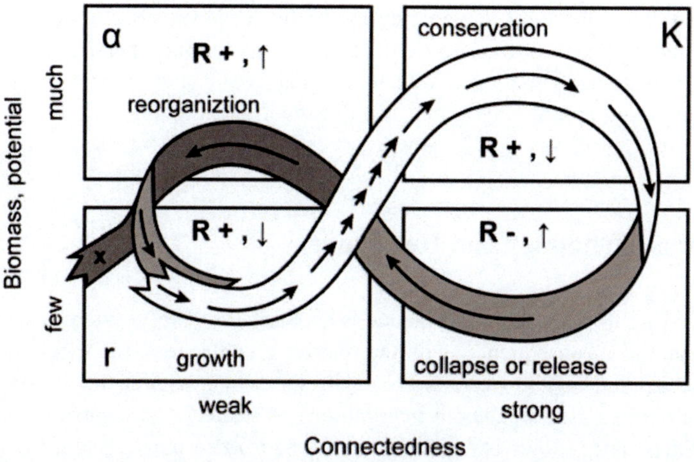

Fig. 6.2 Representation of the adaptive cycle according with Holling (1986, 2001). Four states can be characterized for the system's dynamics, as a result of the internal processes and external influence. In the growth phase, resilience is high (*R*+) but tends to decrease (arrow down) while transiting towards the conservation phase. Here, resilience diminished (*R*−) and continued decreasing transiting to a collapse, presenting low resilience (*R*−) which tend to increase (arrow up) transiting to the reorganization phase where resilience is high (*R*+) and continue its tendency to increasing. After that, the system moves towards the growth phase to complete the cycle (adapted from Holling, 1986)

develops, and the resilience, although high, tends to decrease as it approaches the conservation phase, which would correspond in this case to the carrying capacity. In this phase, the system tends to be stable while maintaining its functions; and its state corresponds to the classic concept of ecological resilience. In this phase, resilience is lower than in the previous phase. Given that resilience dynamically maintains a decreasing tendency, the capacity of the system to respond to disturbances also decreases, making it more vulnerable to change. In this way, the disturbance causes a decrease that, in our example, manifests as a decrease in population abundance. In this state, the system, while maintaining low resilience, tends to enter a reorganization phase that corresponds to so-called engineering resilience (Holling 1973), where resilience continues to increase. In this phase, the system is reorganized to subsequently start the growth process again. In this sense, engineering resilience is related to the efficiency of system functions, while ecological resilience is related to the resilience associated with maintaining existing functions (Gunderson and Holling 2002).

These processes, briefly illustrated with the example of a population, occur for any of the individual systems, or hierarchical levels, illustrated in Fig. 6.1 and even for fishing systems in general. Likewise, the duration, the rate of change, and the vulnerability relative to each of the phases of the adaptive cycle depend on the biological and ecological nature of the system or on the characteristics of the socioeconomic and governance systems. The relevant part for management in this context is the possibility of recognizing the particular phase or state of the adaptive cycle of the exploited system, recognizing its attributes (i.e., persistence, vulnerability, and resilience) in such a way that management and control measures are applied in a more appropriate, timely, and more efficient manner.

Ecosystem Dynamics and Resilience

When referring to ecosystems, resilience is associated with ecosystem structure and function and, as a consequence, with the relative abundance of the species composing it, the strength of the interactions in the food web, as well as the role of the species or species assemblages in maintaining its structure and function, such that knowing the energy flows is a key element for resource management (May 1972, 1973, 1974, Walker et al. 2004, Emmerson and Raffaelli, 2004, Vallina and Le Quéré 2011, Gunderson 2000, Hughes et al. 2005, Llope et al. 2011).

According to Ulanowicz (1986), in terms of the energy flows of an ecosystem, the difference between the development capacity (Eq. 3.2) and the organization of an ecosystem (Eq. 3.1) is defined as system overhead ϕ (Eq. 3.3) or "strength in reserve" expressed as

$$\phi = -\sum_{i,j} T_{ij} \log\left(\frac{T_{ij}^2}{T_{i\circ}T_{\circ j}}\right)$$

This reserve energy represents residual freedom, relative to ascendency, expressed as conditional entropy. This entropy represents the indeterminacy of flows (flows without constraint, sensu Ulanowicz 2009), associated with the multiple flow pathways over which energy could flow, i.e., functional trophic redundancy (Ulanowicz 2004, 2018). To better understand this concept and its relationship with resilience and its usefulness in resource management, consider the marginal diversity of flows in terms of their origin and destination (development capacity sensu Ulanowicz 2018) as:

$$C = -\sum_{i,j} \left(\frac{T_{i,j}}{T_{\circ,\circ}}\right) \log\left(\frac{T_{i,j}}{T_{\circ,\circ}}\right) \tag{6.1}$$

where $\left(\frac{T_{i,j}}{T_{\circ,\circ}}\right)$ represents a posteriori probability of the transfer of flow leaving species (prey) i and entering species (predator) j. Considering only the origin (prey), the expression is represented as:

$$C_{\mathrm{O}} = -\sum_{i}\left(\frac{T_{i,\circ}}{T_{\circ,\circ}}\right)\log\left(\frac{T_{i,\circ}}{T_{\circ,\circ}}\right) \tag{6.2}$$

where $C \geq C_{\mathrm{O}} \geq 0$, and similarly, for the destination of the flows (predators) is represented as:

$$C_{\mathrm{d}} = -\sum_{i}\left(\frac{T_{\circ,j}}{T_{\circ,\circ}}\right)\log\left(\frac{T_{\circ,j}}{T_{\circ,\circ}}\right) \tag{6.3}$$

where $C \geq C_{\mathrm{d}} \geq 0$, and, consequently, the diversity of flows will be greater than the joint diversity of flows, $C_{\mathrm{O}} + C_{\mathrm{d}} \geq C$, and this occurs due to the limitation of flows that defines how the components are connected (order of the system) reducing the indeterminacy. According to this, the difference, $C_{\mathrm{O}} + C_{\mathrm{d}} - C$, measures the flow constraints associated with system topology, and after algebraic manipulation with the previous equations and weighting the total flows, the definition of ascendency expressed in Eq. (3.1) is obtained; as a consequence, $C - A = \phi$, where ϕ, as defined above (or Eq. 3.3), is a measure of the redundancy of flow pathways, or conditional entropy.

According to Ulanowicz (1986), the amounts that define ascendency, development capacity, and overhead are composed of different contributions such as imports, exports, dissipated energy, and flows redundancy. Specifically, contributions to overhead are expressed as:

$$\text{imports} : \phi_{\mathrm{I}} = -\sum_{j=1}^{n} T_{0,j}\log\left(\frac{T_{0,j}^{2}}{T_{0,\circ}T_{\circ,j}}\right)$$

$$\text{exports} : \phi_{\mathrm{E}} = -\sum_{j} T_{i,n+1}\log\left(\frac{T_{i,n+1}^{2}}{T_{i,\circ}T_{\circ,n+1}}\right)$$

$$\text{dissipations(i.e., respiration)} : \phi_{\mathrm{D}} = -\sum_{j} T_{i,n+2}\log\left(\frac{T_{i,n+2}^{2}}{T_{i,\circ}T_{\circ,n+2}}\right)$$

$$\text{redundancy} : \phi_{\mathrm{R}} = -\sum_{i,j=0} T_{i,j}\log\left(\frac{T_{i,j}^{2}}{T_{i,\circ}T_{\circ,j}}\right)$$

where $n + k$ represents destinations other than predators j, and 0 is a different source than prey i.

If we assume that in a given ecosystem, there are no exports or imports, and if we consider that the dissipated energy is unused energy, then the redundancy represents the possibilities of trophic exchange; and in terms of overhead, this represents the conditional entropy associated with this redundancy.

As stated above, overhead is a system's energy in reserve and corresponds to the entropy available in the network, which in turn is represented by the probabilities of non-conditioned flows or the degree of network indeterminacy. Theoretically, in an undisturbed and stable ecosystem, the flows' limitation is equally stable, and a certain degree of order is maintained where the energy balance of the ecosystem, derived from trophic relationships and energy exchanges, is self-sustaining. If one of the components from this system loses biomass due to some disturbance, for example, fishing, several mechanisms come into play that explain the ecosystem's ability to restore itself. Thus, for example, from a population dynamics perspective, the response to biomass loss due to a disturbance is the increase in the production rate that generates surplus production to compensate for the loss. These responses represent a compensatory mechanism for restoring biomass, and the speed of restoration depends on the growth rate as an intrinsic characteristic of the life history of the population. The loss of biomass will affect the predators and prey of the target species through the trophic exchanges defined by the organization of the ecosystem, entering into play the additional distribution of energy through the overhead' redundancy, modifying the probabilities of flows of the initial state. The result of these processes is, at the end, the compensation of the processes altered by the disturbance, allowing the restoration of the population. This would be, in essence, the mechanism that explains the resilience of the trophic network of the ecosystem. Of course, when the disturbances come from different causes (i.e., environment, fishing, etc.) and act on several species simultaneously, the compensatory processes involving resilience would not be easy to identify and isolate, with the joint effects being represented through the probabilities of energy flowing in the entire ecosystem.

According to Ulanowicz (1986, 2009), overhead is the strength in reserve that an ecosystem can use to counter the effects of disturbances and restore itself from impacts; that is, it is directly related to resilience and, to a large extent, as we observed before, corresponds to flow redundancy. But, how does resilience work? Where is this reserve energy found, or where does it come from?

In a balanced undisturbed ecosystem, there would be no surplus production because restoration mechanisms come into play only when populations and ecosystems are disturbed. In this state of balance, the reserve energy exists, but it is not stored in a place ready to be taken and used by the ecosystem when necessary, in the event of a disturbance. The overhead represented as redundancy (energy that does not contribute to the organization of the ecosystem) is a latent energy, whose flow is not conditioned, and exists in each species through its trophic relationships. For simplicity, Fig. 6.3a represents a set of nodes and connections, in which the arrows represent the direction of the flow and its thickness represents the intensity of the flows. The latent energy, which is shown in Fig. 6.3a as the space delimited by the dotted line around each node, represents the overhead which, by definition, corresponds to an unconditioned flow probability; that is, it is the part of the useful energy whose probability of flow is uncertain and that, when there is no disturbance, behaves with the same probability of conditional flow (conditional entropy), but

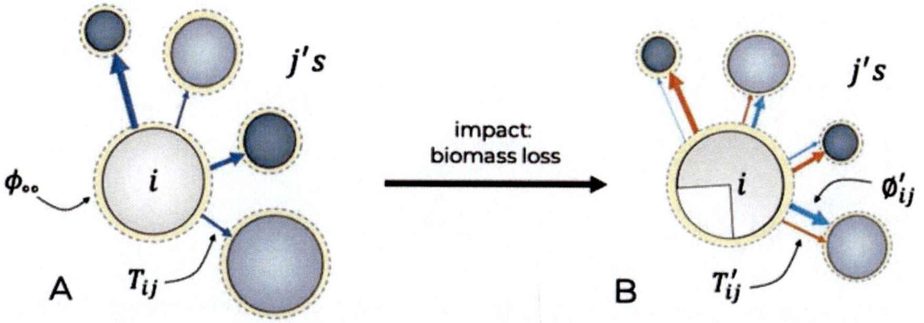

Fig. 6.3 Simplified graphic representation of the relationship between overhead, resilience, and the effect of biomass extraction in an ecosystem. Each species or node in a food web is related to others by exchanging energy through predator (j)–prey (i) relationships. The figure on the left (**a**) shows these relationships. Let us assume that this figure represents part of an undisturbed system. According to Ulanowicz (1986, 2009), the energy flows from a prey to a predator (T_{ij}) are of two types, limited or conditioned flows (black arrows) and unconditioned flows (white arrows). The former is responsible for maintaining the structure and organization of the ecosystem and are expressed as Ascendency; while the seconds correspond to the internal entropy (Overhead, $\varnothing_{\circ\circ}$) and constitute the energy available in the system (food web) to deal with disturbances. In this non-disturbance condition, the probabilities of these flows remain equally stable. When a disturbance occurs, figure to the right (**b**), the probabilities of the conditioned flows maintain the structure and the link between the components, while the internal entropy represents random flow probabilities, where such flows or energy is oriented towards those regions of the trophic network that require more energy to maintain thermodynamic balance. These latter flows, as a whole, constitute the resilient capacity of the ecosystem

that, when a disturbance occurs, its probability changes and is defined by the requirement of the thermodynamic balance maintenance process when it is altered.

When a disturbance produces a reduction in biomass, in addition to the response of the population itself to restore that loss; the probabilities of flows change, and the other associated nodes are affected. In Fig. 6.3b, the central node represents a loss of biomass, the adjacent nodes to which energy flows enter reduce in intensity due to the lower available biomass, and the probabilities of the latent energy change according to the strength of the interactions.

The reserve energy, sensu Ulanowicz, represents the latent free energy (or conditional entropy) that, given the redundancy of flows, is distributed differentially through the energy inputs and/or outputs between species. In this way, the global resilience of the ecosystem comes partly from the energy provided by population processes (surplus production), and partly from the differential distribution of energy towards the other components of the ecosystem, through the flows represented by the prey–predator relationships.

Regarding fishing, where there are no inputs to production, the properties of resilience cannot be directly managed or controlled; however, it is feasible that when knowing their attributes and behavior, management decisions consider the consequences of changes in resilience to take advantage of it and favor adaptability towards the sustainable use of living resources; and, eventually, to improve social welfare and even promote development. Let us briefly review two examples: the effects of climate change in the Campeche Bank and the usefulness of this knowledge to defining harvest rates for the different species within the ecosystem.

Climate and Resilience in the Campeche Bank

In the Campeche Bank, in the southern Gulf of Mexico, effects of climate changes have been demonstrated on several fishing stocks (Arreguín-Sánchez 2012, 2019, Arreguín-Sánchez et al. 2015, 2017, 2019). Figure 6.4 shows anomalies of the Atlantic Multidecadal Oscillation, AMO, index, which is mainly based on temperature. Using the trophic model of the ecosystem, Ecosim (Arreguin-Sanchez et al. 2004, 2008), the effect of the AMO was represented for the period from 1956 to 2010 by forcing the production rate of the primary producers, and allowing the signal propagates across the trophic network. The simulation aims to emulate the possible historical effect of climate change in accordance with the above-mentioned references. The resilience of the ecosystem was computed (Arreguín-Sánchez 2014) for each year through the relationship

$$\phi = -\sum_{i,j} \log\left(\frac{T_{ij}^2}{T_{i\circ}T_{\circ j}}\right)$$

and the changes in the entropy of the ecosystem were estimated as

$$\mathcal{E} = 1 - ({}^{A}\!/_{C})$$

Figure 6.4 shows that ecosystem resilience is lower at the time of the regime shift, while entropy shows greater uncertainty for the same period. In a precautionary sense, what these trends suggest is that at the time of the regime shift, the ecosystem was more vulnerable to potential impacts, and as a consequence, in terms of management, it is the time when compliance of control measures should be strongly monitored to allow the ecosystem to naturally restore itself and avoid undesirable overfishing conditions. There are two relevant aspects to be mentioned: (1) the time lag between the regime shift, shown by the trends of change in the AMO‘s anomaly, and the change in the resilience. In this case, a lag-time of 5 years corresponds to the ecosystem resistance; and (2) the restoration of the ecosystem, which consist in its dynamic reconfiguration over time, as a consequence of its adaptability, without necessarily recovering its previous configuration. This case will be revisited with more detail in Chap. 10.

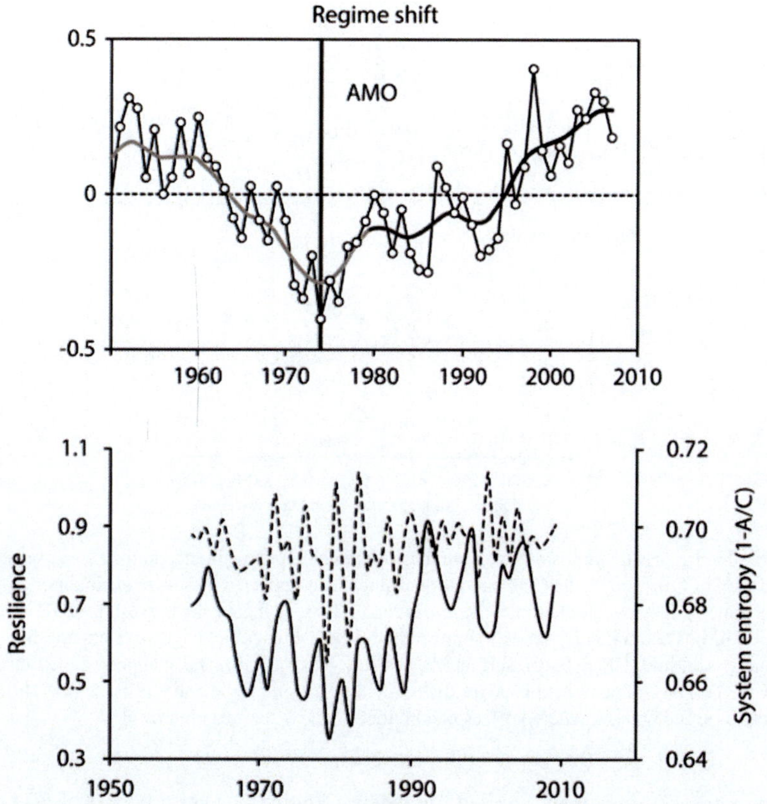

Fig. 6.4 Changes in resilience related to regime shift in the Campeche Bank, southern Gulf of Mexico. In the picture on top, changes in the Atlantic Meridional Oscillation indicating the time of regime shift can be observed, showing the end of a cooling period and the beginning of a warming tendency. Picture on the bottom shows two resilience estimations, the system entropy derived from the Ulanowicz (1986, 2009) theory, in the dashed line; and that proposed by Arreguin-Sanchez et al. (2004) based on supply and demand of energy in the trophic network (solid line). Both the indices respond to regime shift impact, one increasing uncertainty (dashed line) and the other by decreasing resilience (solid line). In both the cases, these trends can be used as base for precautionary management

Fishing and Resilience Threshold: The Gulf of California

The second example corresponds to the central west region of the Gulf of California ecosystem. Based on a trophic model of the ecosystem (Arreguín-Sánchez et al. 2021), the effect of biomass extraction on ecosystem resilience was estimated by computing the harvest rate for each functional group that corresponds to the threshold reference level after which the ecosystem changes its organization and functioning; that is, a limit reference point referred to the ecosystem. For these estimates, the ecosystem was brought to a state of no-fishing, and subsequently, for each exploited

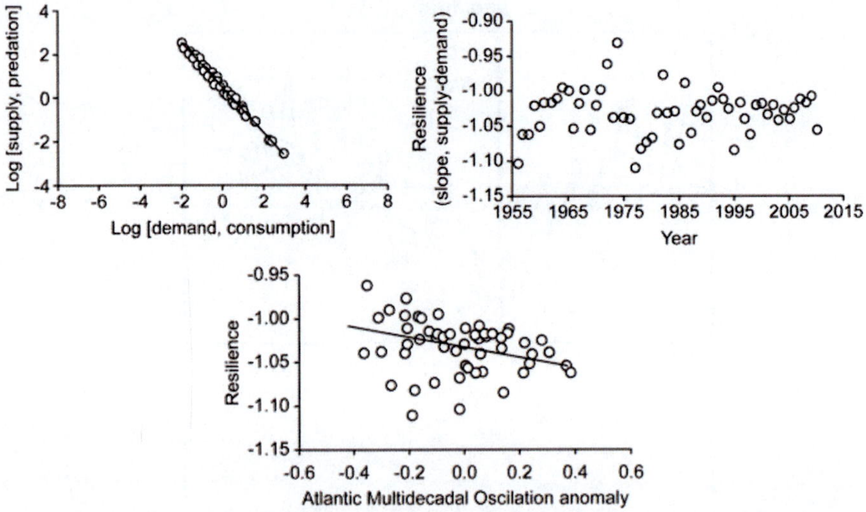

Fig. 6.5 Figure on top, trends of the anomalies of the change in resilience in the ecosystem of the central Gulf of California, when some individual stocks are impacted by severe exploitation. Trends were obtained under the simulation of the impact of a gradually increasing harvest rate (HR = catch/biomass ratio). The limit harvest rate to impact the resilient capacity of the ecosystem are shown for clupeoids, solid line (HR = 0.40), shrimp, dashed line (HR = 0.36), and snappers, dotted thin line (HR = 0.34). The isoline of resilience for the limit exploitation for the entire ecosystem is shown in the picture on the bottom, where species can be identified by their trophic level

group, a harvest rates were applied, increasing gradually over a period of 50 years. For each species and year, resilience was represented as the overhead, sensu Ulanowicz (1986). Taken the anomaly of the time series of resilience, the inflection point was considered as the point of no return, such that the harvest rate corresponding to such inflection point represents the critical harvest threshold, after which, the organization and functioning of the ecosystem would be affected due to biomass extracted from that group in particular (Fig. 6.5). When arriving at these estimates for all groups in the ecosystem, a graph was obtained, in which all species are represented by their trophic level and the isoline represents the upper limit of harvest rates that could be applied on the species in the ecosystem so that its resilience is not affected by the loss of biomass.

Of course, the concepts outlined above are of interest for management when it is perceived as a necessity to maintain the resilience of ecosystems, especially in a precautionary context. For example, knowing trends in resilience to change, and the critical limits for biomass extraction can support preventive criteria. Suppose, for example, the convenience of having a more refined fishing control system at times like in which the regime shift occurred, when the system could have been more vulnerable, such as that happened in the case of the southern Gulf of Mexico illustrated in Fig. 6.4. In this case, some fisheries collapses can be easily associated with the absence of a process such as this (see Chap. 10).

References

Arreguín-Sánchez, F. 2012. The dynamics linking biological hierarchies, fish stocks and ecosystems: implications for fisheries management. In *Models of the ecological hierarchy: from molecules to the ecosphere*, ed. F. Jordán and S.E. Jørgensen, 501–516. Oxford, UK: Elsevier B.V.

———. 2014. Measuring resilience in aquatic trophic networks from supply–demand-of-energy relationships. *Ecological Modelling* 272: 271–276.

———. 2019. Climate change and the rise of the octopus fishery in the Campeche Bank, Mexico. *Regional Studies in Marine Science* 32: 100852.

Arreguin-Sanchez, F., M. Zetina-Rejón, S. Manickchand-Heileman, M. Ramırez-Rodrıguez, and L. Vidal. 2004. Simulated response to harvesting strategies in an exploited ecosystem in the southwestern Gulf of Mexico. *Ecological Modelling* 172 (2–4): 421–432.

Arreguin-Sanchez, F., M. Zetina-Rejón, and M. Ramírez-Rodríguez. 2008. Exploring ecosystem-based harvesting strategies to recover the collapsed pink shrimp (Farfantepenaeus duorarum) fishery in the southern Gulf of Mexico. *Ecological Modelling* 214 (2–4): 83–94.

Arreguín-Sánchez, F., P. del Monte-Luna, and M.J. Zetina-Rejón. 2015. Climate change effects on aquatic ecosystems and the challenge for fishery management: pink shrimp of the Southern Gulf of Mexico. *Fisheries* 40 (1): 15–19.

Arreguín-Sánchez, F., P. del Monte Luna, M.J. Zetina-Rejón, A. Tripp-Valdez, M.O. Albañez-Lucero, and T.M. Ruiz-Barreiro. 2017. Building an ecosystems-type fisheries management approach for the Campeche Bank, subarea in the Gulf of Mexico Large Marine Ecosystem. *Environmental Development* 22: 143–149.

Arreguín-Sánchez, F., M.O. Albañez-Lucero, P. Del Monte-Luna, and M.-J. Zetina-Rejón. 2019. Fishery resource management challenges facing climate change. In *Mexican aquatic environments*, ed. A.L. Ibañez, 181–194. New York, NY: Springer.

Arreguín-Sánchez, F., P. del Monte-Luna, M.O. Albañez-Lucero, M.J. Zetina-Rejón, A. Tripp-Quezada, T.M. Ruiz-Barreiro, and J.C. Hernández-Padilla. 2021. How much biomass must remain in the sea after fishing to preserve ecosystem functioning? The case of the sardine fishery in the Gulf of California, Mexico. In *Marine coastal ecosystems modelling and conservation*, ed. M. Ortiz and F. Jordán, 143–161. Cham, Switzerland: Springer.

Cushing, D.H. 1975. *Marine ecology and fisheries*. Cambridge, UK: Cambridge University Press.

———. 1995. *Population production and regulation in the sea: a fisheries perspective*. Cambridge, UK: Cambridge University Press.

Emmerson, M.C., and D. Raffaelli. 2004. Predator–prey body size, interaction strength and the stability of a real food web. *Journal of Animal Ecology* 73 (3): 399–409.

Folke, C. 2006. Resilience: the emergence of a perspective for social–ecological systems analyses. *Global Environmental Change* 16: 253–267.

Folke, C., S.R. Carpenter, B.H. Walker, M. Scheffer, T. Elmqvist, L.H. Gunderson, and C.S. Holling. 2004. Regime shifts, resilience and biodiversity in ecosystem management. *Annual Review in Ecology, Evolution and Systematics* 35: 557–581.

Folke, C., S.R. Carpenter, B. Walker, M. Scheffer, T. Chapin, and J. Rockström. 2010. Resilience thinking: integrating resilience, adaptability and transformability. *Ecology and Society* 15 (4): 20.

Gunderson, L.H. 2000. Ecological resilience—in theory and application. *Annual Review of Ecology and Systematics* 31 (1): 425–439.

Gunderson, L.H., and C.S. Holling, eds. 2002. *Panarchy: understanding transformations in human and natural systems*. Washington, DC: Island Press.

Herrera, G., and G. Rodríguez. 2016. Resiliencia y turismo: El caso de la ciudad de Baños de Agua Santa-Ecuador. *HOLOS* 3: 229–250.

Holling, C.S. 1973. Resilience and stability of ecological systems. *Annual Review of Ecology and Systematics* 4 (1): 1–23.

———. 1986. Resilience of ecosystems; local surprise and global change. In *Sustainable development of the biosphere*, ed. W.C. Clark and R.E. Munn, 292–317. Cambridge, UK: Cambridge University Press.

———. 2001. Understanding the complexity of economic, ecological, and social systems. *Ecosystems* 4 (5): 390–405.

Hughes, T.P., D.R. Bellwood, C. Folke, R.S. Steneck, and J. Wilson. 2005. New paradigms for supporting the resilience of marine ecosystems. *Trends in Ecology & Evolution* 20 (7): 380–386.

Llope, M., G.M. Daskalov, T.A. Rouyer, V. Mihneva, and CHAN, K. S., Grishin, A. N., & Stenseth, N. C. 2011. Overfishing of top predators eroded the resilience of the Black Sea system regardless of the climate and anthropogenic conditions. *Global Change Biology* 17 (3): 1251–1265.

May, R.M. 1972. Will a large complex system be stable? *Nature* 238 (5364): 413–414.

———. 1973. Qualitative stability in model ecosystems. *Ecology* 54 (3): 638–641.

———. 1974. *Stability and complexity in model ecosystems*. Princeton, NJ: Princeton University Press, 265 p.

Ulanowicz, R. 1986. *Growth and development: ecosystems phenomenology*. New York, NY: Springer, 203 p.

Ulanowicz, R.E. 2004. Quantitative methods for ecological network analysis. *Computational Biology and Chemistry* 28 (5–6): 321–339.

Ulanowicz, R. 2009. The dual nature of ecosystem dynamics. *Ecological Modelling* 220: 1886–1892.

Ulanowicz, R.E. 2018. Biodiversity, functional redundancy and system stability: subtle connections. *Journal of the Royal Society Interface* 15 (147): 20180367.

Vallina, S.M., and C. Le Quéré. 2011. Stability of complex food webs: resilience, resistance and the average interaction strength. *Journal of Theoretical Biology* 272 (1): 160–173.

Walker, B., C.S. Holling, S.R. Carpenter, and A. Kinzig. 2004. Resilience, adaptability and transformability in social–ecological systems. *Ecology and Society* 9 (2): 5.

Chapter 7
Stock and Ecosystem (Holistic) Biological Reference Points

Population-Based Biological Reference Points

Biological reference points, BRPs (Caddy and Mahon 1996), are performance indicators of exploited populations, they refer to fishing level or harvest rate values and are useful for management purposes because they provide a clear definition of the state of exploitation of a resource and, as a consequence, allow a tangible basis for managing fisheries and take decision process. The virtue of these reference points is that they are independent of the intrinsic characteristics of the exploited populations, and therefore, their interpretation has the same conceptual basis for any fish resource and life history and also is of easy understanding for managers and fishermen. Different BRPs have been proposed or have appeared in the literature at various times over several decades, for example, maximum sustainable yield (Schaefer 1954), marginal yield (Gulland and Boerema 1973), exploitation rate at a given level of spawning biomass (Brooks et al. 2008), maximum economic yield (Clark 1990), and harvest rate that maximizes yield per recruit (Katsukawa 2005), among others. In all cases, the BRP must be defined in terms of the fishing effort, fishing mortality, harvestable biomass, or harvest rate and, although these indicators can generally be available, in practice, those managers responsible for making decisions should decide the BRP to be applied, according to the particular circumstances of each fishery.

BRPs are very useful and intuitive tools because the measures or control rules derived from them are tangible, observable, and easily interpreted by the fisheries sector. In general, there are two types of BRPs: target, which refers to the harvest rates aimed at the long-term sustainability of the exploited resource, and limit, which is a threshold that indicates risky conditions for the exploited fishing resources, beyond which their exploitation ceases to be sustainable, causing deterioration and collapses of fisheries. Of course, what is desirable in fisheries management is for managers to make decisions that will avoid the risk of falling below the BRP limit, entering the precautionary measures into play (Gabriel and Mace 1999).

F. Arreguín-Sánchez, *Holistic Approach to Ecosystem-Based Fisheries Management*, https://doi.org/10.1007/978-3-030-96847-2_7

The use and application of BRPs as management tools is not new, although the set of criteria was not always used systematically, as tends to happen in recent years, but rather they were developed and applied to particular exploitation state and needs of each resource. Towards the 1950s and later, the prevailing regulating criteria generally referred to maximizing conditions, whether they were fishing yields, net income, jobs, among others, or even multiple optimization criteria (Mardle and Pascoe 1999). However, as demonstrated later, these strategies tend to cause the overexploitation of exploited populations (Hilborn 2007). Currently, the general concept is to use a BRP as a criterion to maintain sustainable fishing considering that the availability and productive capacity of populations vary over time and that the extraction capacities of fisheries require adaptability strategies for that purpose. This implies, in general, the desirability and need to have a set of BRPs that facilitate sustainable management conditions. The usefulness of BRPs has been reflected, in recent decades, in their applicability to generic guidelines, such as those in the United Nations Fish Stocks Agreement (Munro 2000, Balton and Koehler 2010) and the Code of Conduct for Responsible Fisheries (FAO 1995) and those created by certifying agencies such as the Marine Stewardship Council (MSC, https://www.msc.org), among others.

Concerning an ecosystem, population BRPs are useful for controlling the retained biomass or the fishing mortality. As a principle, the criteria defined by population BRPs work because controls on extraction tend to maintain the production capacity of populations within the limits that allow populations and ecosystems to restore extracted biomass per unit of time. In this context, in a theoretical sense, population BRPs are fully compatible with ecosystem-based analysis criteria. However, this is a perspective that does not properly consider the processes associated with the interdependencies among populations in ecosystems; for example, if we choose as BRP target the maximum sustained yield or the marginal yield ($F_{0.1}$), it would not be possible to bring all the species to that level simultaneously because energy balances are not linear between or within the trophic levels; and even more, they do not consider differences between life histories and the consequences when stocks respond differently interactions and to the perturbations. Consequently, it is highly recommended to also consider holistic attributes of the ecosystem. The relevant aspect is maintaining the sustainability of fishing in the two interacting but also interdependent biological hierarchies.

The main problem regarding regulatory effectiveness when using BRPs lies more in failed implementation than in concept. This is probably due to the entrenched custom of two conventional assumptions of population models that were used, justifiably, for several decades: (1) the stability of populations and therefore of ecosystems and (2) the concept of fishing as the unique or main source of variation of the trends of the stock's abundances. This last concept does not imply that the effects or contribution of the environment on population variability were not recognized; for example, the above assumption refers to the fact that significant changes in population abundance trends mainly due to fishing and that environment does not contribute with long-term trend, but to the short-term variability. In this scenario, management measures tend to remain static; resulting, at the present, in an inability to adapt these measures to changes in populations over time and even less responding to the reconfiguration of ecosystems.

Ecosystem-Based Biological Reference Points

As mentioned in Chap. 5, and considering an ecosystem as a representation of the dynamics of a trophic network, holistic concept refers to indicators of network performance that reflect attributes such as organization, functioning, cohesion, and resilience (Ulanowicz 1986, 2009, 2018), among others; which can be used, according to their properties, as references to define ecosystem states. Regarding an exploited ecosystem, in particular regarding fishing, it is easy to understand that biomass extraction affects the energy transfer between species. As shown in the previous chapter, biomass extraction has two response levels, one at the population level, generating surplus production, and the other, at the ecosystem level, through adjustments in energy exchange between populations, determined by conditioned flows of energy that maintain the order of an ecosystem (ascendency), and the conditioned entropy or redundant flows, allowing energy to be redirected towards maintaining a system's thermodynamic balance.

Formally, the deterioration of an ecosystem can be measured by the entropy generated in it, affecting conditioned flows and, as a consequence, its organization and functioning. Concerning the redundant flows, or conditioned entropy, they are part of the ecosystem dynamics that allows the ecosystem to react to perturbation. As mentioned before, as long as the system is not disturbed, this conditioned entropy or latent energy flows through the network with the same probabilities as the conditional flows responsible to maintain, dynamically, the ecosystem organization, i.e., the thermodynamic balance. However, when the system is disturbed, such as by removing biomass, while the probabilities of the conditioned flows remain and tend to maintain the order, the probabilities of the redundant flows (conditioned entropy) are modified, tending to compensate for the disturbance. These flows are entropy that flows throughout the network topology; therefore, their probability responds to the needs of maintaining the balance of the system, modifying the flow probabilities accordingly. Obviously, if the disturbance is small, the surplus production of the population compensates its biomass, and the flow probabilities compensate for changes in the distribution of energy in the ecosystem, and the balance is recovered. In this context, the measurement of the entropy generated by the probability changes in the redundant flows would be a measure of the disturbance or the degree of deterioration of the ecosystem and, of course, would have a limit defined by the production capacity of the populations (as source of energy) as well as of the capacity and amount of redundant flows used to redistribute the energy that compensate for the thermodynamic unbalance of the ecosystem. This capacity, of course, has a limit that is given by the amount of the available conditioned entropy or latent energy in the system (overhead sensu Ulanowicz 1986). This limit, expressed by the change of the trend of conditioned entropy, given a continuous and increasing disturbance, would be the limit of the ecosystem after which it is no longer possible to restore the previous balance, thus leading to another state of the ecosystem. In this sense, it should be noted that the restoration capacity is not a timeless attribute but takes time, depending on the intrinsic characteristics of the food web. In the case of fishing, this limit, expressed as the change in entropy trends, could be identified by simulation experiments and expressed through the proportion of biomass retained by the fishing regarding the availability of biomass of the population; that is, by the harvest rate. If this is identified for different populations, the limits for an ecosystem, as holistic reference points, can be determined. This is the concept developed by Arreguín-Sánchez et al. (2022a) and proposed as an ecosystem-based biological reference point (EBRP), when referring to a particular resource; or as the noxicline, when referring to the entire ecosystem. Figure 7.1a shows the EBRPs for several individual species, while Fig. 7.1b shows noxicline.

As mentioned in previous chapters, the environment imposes conditions that do not allow assumptions of nature's stability to be currently acceptable. The response of populations to a changing environment leads them to adapt and ecosystems to reconfigure. The response mechanisms to a disturbance such as fishing add to this process of reconfiguration of natural systems, leading to a continuous adjustment of thermodynamic balances, which leads to alternative ecosystem states. These processes are dynamic and are continuous, and the limits of adaptability and reconfigurability are defined by the attributes of the different hierarchies of biological organization, whether they are characteristics of the life histories of the species reflected at the level of individuals, such as metabolism; or for the populations, expressed as reproductive success and survival; and of ecosystems, e.g., restoring energy flows, expressed as a reconfiguration of ecosystem organization and functioning. Under these conditions, when a medium- or long-term trend manifests, such as occurs with the effect of climate; and when, in addition, ecosystems are disturbed by extracting biomass, as is the case of fishing, the processes described above converge. However, under these conditions, assuming no overfishing, restoration to the initial state of the ecosystem will most likely not occur, as natural consequence of the reconfiguration caused by climatic forcing; and when a climate event is highly intense, even without fishing, the ecosystem would not necessarily return to its initial state. In this context, adjustments by the populations and the thermodynamic balance of the ecosystem are continuous dynamic processes that will necessarily require adaptation strategies, also continuous, for the management strategies; a condition for which the determination of EBRPs and the corresponding harvest rates will be very useful.

Change in Climate Patterns and Ecosystem Reconfiguration

One important aspect to consider in the management of living marine resources is that the reconfiguration of ecosystems, induced by climate effects, leads to alternative states and not necessarily to the restoration of the initial state of the ecosystem.

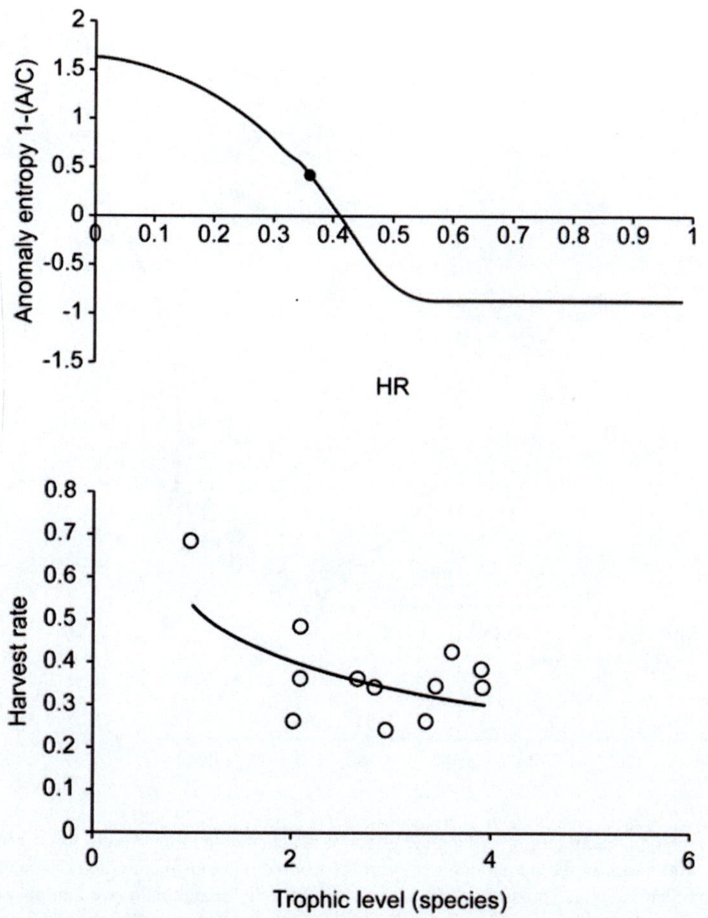

Fig. 7.1 Estimation of the noxicline (Arreguín-Sánchez et al. 2022a) for the ecosystem of northern Gulf of California. The top picture represents the simulation of the impact of biomass retention of an individual functional group or species in the ecosystem represented by a trophic network. Harvest rate (HR = Catch/Biomass ratio) was gradually increased and the effect on the entropy, given as $\epsilon = 1 - (A/C)$ was computed (A represents the Ascendency and C the Capacity of development sensu Ulanowicz (1986, 2009). The inflection point of the anomaly of entropy is assumed as the limit of no return and as consequence the limit harvest rate. This simulation can be developed for each component of the ecosystem and related with the respective trophic level resulting in an isoline named noxicline (*picture below*) and represents the limit harvest of the entire ecosystem

Consider Fig. 7.2, and the historical trend for the anomaly of temperature, as representative of the environmental forcing on ecosystems. During the stable period, approximately from 1930 to 1980, the climate induced trendless variability. For this period, given the relative stability, it is possible to assume that in the face of any disturbance, the system tends to restore the ecosystem to its prior state, provided that the critical thresholds of the ecosystem itself are not exceeded. When the climate

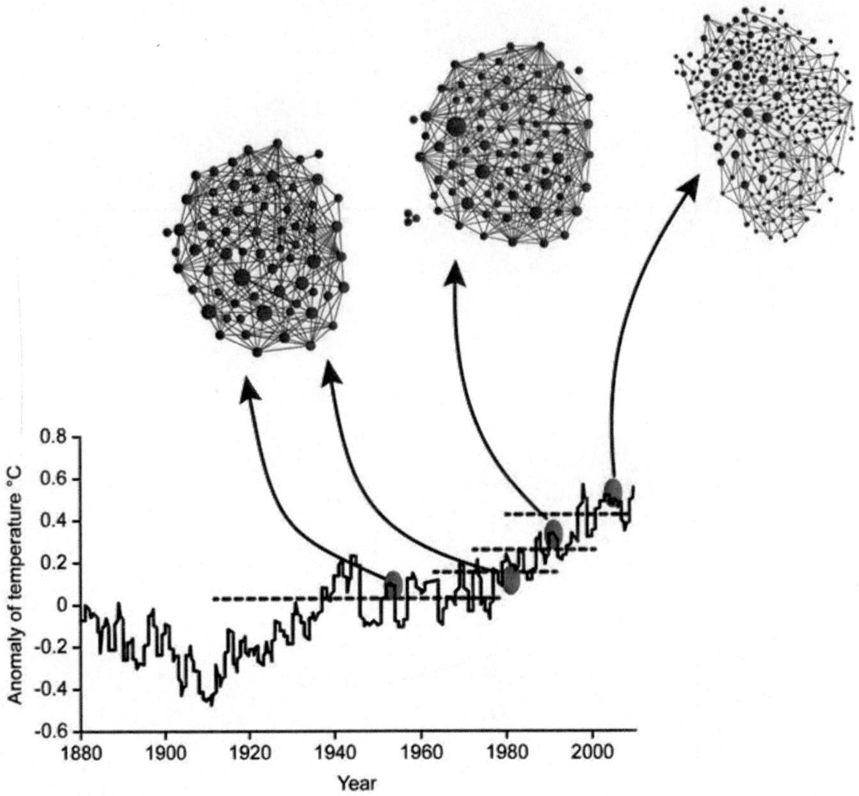

Fig. 7.2 Diagram attempting to explain the effect of the change in climate patterns on ecosystem' dynamics. Networks on the upper side represent the trophic relationships in the ecosystem. When the environment is stable, ecosystem remains also reasonably stable; then, we can also expect a stable carrying capacity as some population dynamics models assume; and some assumptions like the fishing is the main driver of stock abundance can be reasonably taken; or that the collapsed stocks can recover previous sizes when some measures are applied can also be reasonably assumed. When climate pattern changes with tendency over time, the metabolism at individual-fish level is affected, populations adapt and ecosystems reconfigure. In the diagram, the ecosystem on right have the same components as that on left, but its organization and functioning is not necessarily the same, being attributes like, for example, the resilience, different

trend forces the system in the medium or long terms, the ecosystem tends to reconfigure itself continuously, which is represented by the period from 1980 onwards. Given this process, it would not be expected that the restoration of the ecosystem, as a response to a disturbance that occurred in the present time, would lead to an ecosystem state equal to that of the stable period. This concept is essential because conventional management schemes assume, explicitly or implicitly, stability under which the management schemes do not seem to work, or even worst, increases the risks of undesired disturbance or ungovernability. Two aspects of this concept stand out; first, the need for the adaptability of management strategies to the natural trends of ecosystems and populations; and second, the need for eradicating the view that an ecosystem in an alternative state represents a negative scenario regarding the current one, especially if this alternative state is a consequence of a process of the ecosystem natural evolution.

Catch-Based Ecosystem Reference Points

A possible target biological reference point, under the ecosystem approach, is the harvest rate that maximizes the catch per unit of biomass, $\left(\frac{C}{B}\right)_{\max}$ (Arreguín-Sánchez et al. 2022b). This EBRP corresponds to a concept based on the production capacity of populations in the explicit presence of biological interdependencies represented by the energy flows of the ecosystem. The concept is essentially simple; from the simulation experiment initially described, the ratio $\frac{C}{B}$ can be estimated for harvest rates that increase progressively across time. With this, a vector of values can be obtained where a maximum for this ratio, $\frac{C}{B}$, can be observed, as illustrated in Fig. 7.3. As in the previous case, simultaneously maximizing $\frac{C}{B}$ for all species is not possible; therefore, the management strategy should define specific criteria to prioritize the allocation of harvest rates.

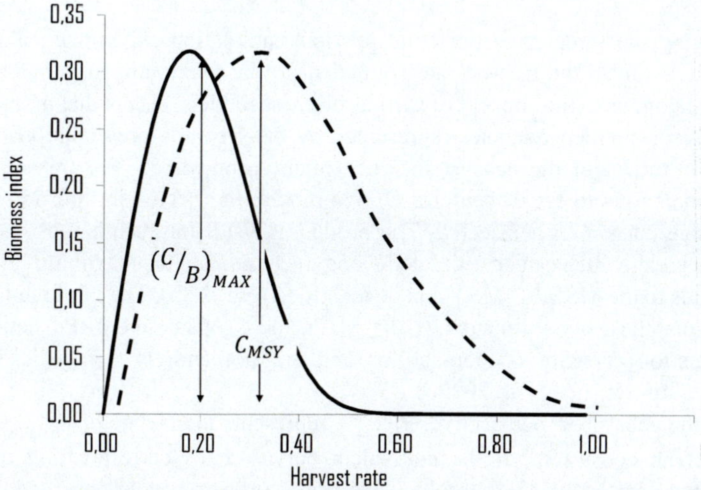

Fig. 7.3 Harvest rates (HR) for the $\left(\frac{C}{B}\right)_{\max}$ ratio, as indicator of maximum productivity of the stock, and the catch at maximum sustainable yield, C_{MSY}; for sharks' resource of the central Gulf of California ecosystem. Estimations were $\text{HR}_{C/B} = 0.18$, and $\text{HR}_{\text{MSY}} = 0.38$, respectively. Both the amounts were estimated through simulation based on a trophic ecosystem model where harvest rate was gradually increased over time. Note that although curves in the figure come from an ecosystem trophic model, it is not properly an ecosystem-based estimation since is only using production capacity information and does not come from holistic indicators (adapted from Arreguín-Sánchez et al. 2022b)

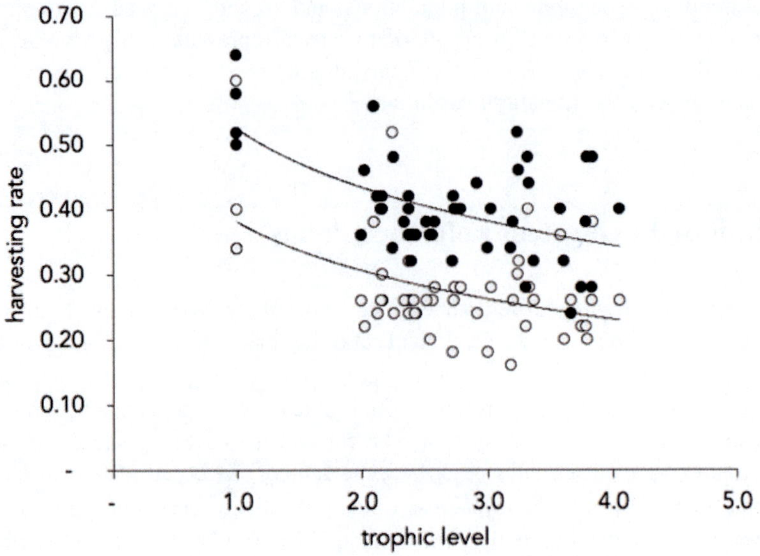

Fig. 7.4 Isolines for $\left(\frac{C}{B}\right)_{max}$ (based on white dots) representing the harvest rate at maximum productivity for fishing; and the noxicline (limit harvest rates after which ecosystem is perturbed in its organization, given from black dots). Each dot represents a species or functional group in the ecosystem of the northern Gulf of California (adapted from Arreguín-Sánchez et al. 2022a, 2022b).

Following the same criteria for the previous simulation experiment, it is also possible to estimate the harvest rate equivalent to the maximum sustainable yield (MSY), analogous to the concept of surplus biomass models, except that in this case, the biological interdependencies represented by the predator–prey relationships of the trophic model of the ecosystem are explicitly considered. For this and in a manner analogous to the Scheafer's (1954) model, the decreasing portion of the ratio change curve *C*/*B* is selected. The product (*C*/*B*)·*B* for each harvest rate will reflect the yield, which can be estimated along the trend, and the maximum level that corresponds to the MSY, C_MSY can be identified (Fig. 7.3b). The important aspect to note is that these two estimators, (*C*/*B*)_MAX and *C*_MSY, like BRPs, depend on the production capacity of populations and are not entirely a holistic indicators (Arreguín-Sánchez et al. 2022b).

From the ecosystem perspective, Fig. 7.4 represents the isoline of $\left(\frac{C}{B}\right)_{max}$ for all species of an ecosystem. If the equivalent harvest rates corresponding to each species, represented by their trophic level, are compared, it is observed that in general, the rates are lower than those corresponding to the noxicline. This difference occurs because the noxicline corresponds to an EBRP limit, given that its definition corresponds to a level after which biomass extraction deteriorates the ecosystem without the possibility of restoration; while the isoline of $\left(\frac{C}{B}\right)_{max}$ would correspond more to an EBRP target if the aim was to maximize productivity.

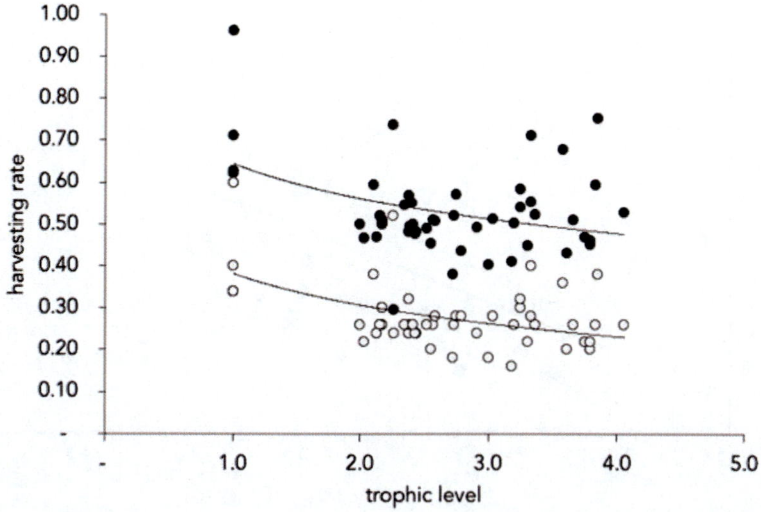

Fig. 7.5 Isolines for $\left(\frac{C}{B}\right)_{max}$ (white dots) representing the harvest rate at maximum productivity for fishing; and the C_{MSY} as the catch at maximum sustainable yield, for the ecosystem of the northern Gulf of California. Each point represents a species or functional group in the ecosystem (adapted from Arreguín-Sánchez et al. 2022b)

Figure 7.5 shows the isoline that would correspond to the MSY for populations, sensu Schaefer (1954), that is, the maximum production capacity of a given population corresponding to 50% of the system's carrying capacity (Arreguín-Sánchez et al. 2022b). For the construction of this reference isoline, the harvest rate corresponding to a ratio $\frac{C}{B} = 0.5$ was directly estimated for each species. Although this yield level is not suggested as a target, it is important to consider it in comparison regarding the noxicline and $\left(\frac{C}{B}\right)_{max}$. Thus, for example, for the same species selected in this example for its trophic level, the harvest rate, HR, identified for the three EBRPs exhibits the following pattern: $HR_{(C/B)_{MAX}} < HR_{noxicline} < HR_{C_{MSY}}$. This pattern is relevant when compared to the concepts rooted in conventional management based on the concept of MSY in independent populations.

Ecosystem's Balanced Surplus-Production

On the other hand, Garcia et al. (2012), after performing simulation experiments with several trophic ecosystem models, suggest that the maximum harvest rate that should be applied in an ecosystem should not exceed 40% of the available biomass. This concept, however, has received criticism (Burgess et al. 2016; Froese et al. 2016), mainly due to the suggestion that to reduce ecosystem impacts, the widest possible range of populations and sizes should be captured in the ecosystem in proportion to their natural productivity; other authors propose that this concept can

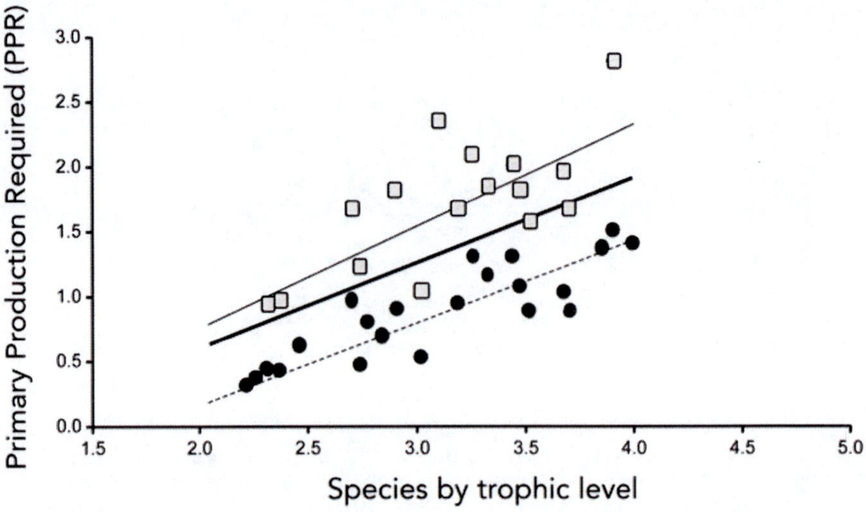

Fig. 7.6 The primary production required to sustain consumption, PPRQ (white dots) and to sustain fisheries, $PPRC_a$ (black dots upper line) for the ecosystem of the central Gulf of California. The theoretical assumption is that the fishery must use the resources in the ecosystem in the same way than nature, that is, as if it were one more predator, represented by gray dots (line in the middle). Then the resulted line is that with gray dots in the middle. Note that the slope of the PPRQ and $PPRC_b$ is the same separated by the surplus production resulted of the exploitation of each stock represents by the trophic level. Also note that the slope of $PPRC_a$ is higher than $PPRC_b$. This means that top predator is being exploited at levels higher than defined by the natural balance expressed by the slope of line PPRQ (adapted from Arreguín-Sánchez et al. 2022b)

be useful and should be explored in greater detail (Jacobsen et al. 2013; Breen et al. 2016). Arreguín-Sánchez et al. (2022c) propose an analogous concept to that of balanced harvesting: fishing should use ecosystem resources in the same way that nature does in terms of the natural mortality rate, that is, as if they were just another predator acting at different trophic levels. Concept defined as balanced surplus production. For this, the authors suggest comparing the primary production required (PPR) to support consumption due to predation PPR_{cons} with the PPR to support catches PPR_{catch}. Figure 7.6 shows these relationships, where PPR_{cons} and PPR_{catch} are provided for the different species in the ecosystem, based on their trophic levels. The application of the concept implies that the slopes of both relationships should be similar. In the example shown, the slope corresponds to $PPR_{catch} > PPR_{cons}$, which implies overfishing higher trophic levels. Arreguín-Sánchez et al. (2022c) show other patterns in these relationships for more than 70 marine and coastal ecosystems. To identify the target catches under this criterion, the slope for the PPR_{catch} should be similar to the corresponding slope for PPR_{cons}; then, from this definition, it is possible to determine the harvest rates to be applied for the different species in the ecosystem, as shown in Fig. 7.7, in which the trend shows clear biological logic if considering the approximate life histories of the different species according to the trophic level they occupy. In this way, the allowable harvest rates would be clearly

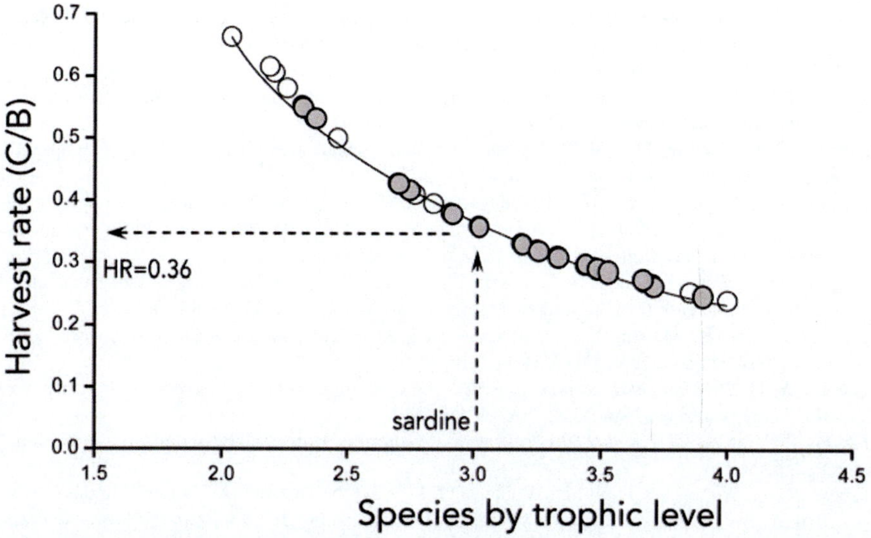

Fig. 7.7 Harvest rates for balance catch (between production capacity and extraction of biomass) along the trophic levels. Note, as expected, that harvest rates for low trophic level are higher than for top predators. Example correspond to the central Gulf of California. (Adapted from Arreguín-Sánchez et al. 2017)

higher for the low-trophic level target species that tend to be *r*-strategists than for high-trophic level species that tend to be *K*-strategists. One aspect of interest, in Fig. 7.6, is that the sum of the biomasses corresponding to each trophic level between the two parallel lines (where the slopes of PPR_{catch} and PPR_{cons} are equal) corresponds to the production capacity (or surplus production) of the ecosystem but not necessarily to the biomass available for fishing. In this context, it is also important to consider that the concept of balanced harvest and the derived harvest rate for each exploited species do not correspond to a holistic ecosystem criterion but only to the production capacity of the species according to their trophic level and their direct and indirect trophic interdependencies.

References

Arreguín-Sánchez F., P. del Monte-Luna, M.J. Zetina-Rejón, M.O., Albáñez-Lucero. 2017. The Gulf of California Large Marine Ecosystem: Fisheries and other natural resources. *Environmental Development* 22: 71–77.

Arreguín-Sánchez F., P. del Monte-Luna, M.J. Zetina-Rejón, L.O. Duarte, D. Lercari, M. Riofrío-Lazo, M.O. Albáñez-Lucero, Danilo Calliari, T. M. Ruiz-Barrerirо, R. Vögler. 2022a. Towards an ecosystem holistic approach to estimate limit harvesting rates.

———. M.J. Zetina-Rejón, P. del Monte-Luna, D. Lercari, L.O. Duarte, R. Vogler, M.O. Albañez-Lucero, D. Calliari, T.M. Ruiz-Barreiro, A. Tripp-Valdez. 2022b. Exploring the estimation of harvest rates for target catch levels from trophic web information.

———. M.J. Zetina-Rejón, P. del Monte-Luna, M.O. Albáñez-Lucero, L.O. Duarte, D. Lercari, D. Calliari, R. Vogler. 2022c. Reference harvest rates estimated from trophic level and production capacity of exploited stocks.

Balton, D.A., and H.R. Koehler. 2010. Reviewing the United Nations fish stocks treaty. *Sustainable Development Law & Policy* 7 (1): 4.

Breen, M., N. Graham, M. Pol, P. He, D. Reid, and P. Suuronen. 2016. Selective fishing and balanced harvesting. *Fisheries Research* 184: 2–8.

Brooks, E.N., K.W. Shertzer, T. Gedamke, and D.S. Vaughan. 2008. Stock assessment of protogynous fish: evaluating measures of spawning biomass used to estimate biological reference points. *Fishery Bulletin* 106 (1): 12–23.

Burgess, M.G., F.K. Diekert, N.S. Jacobsen, K.H. Andersen, and S.D. Gaines. 2016. Remaining questions in the case for balanced harvesting. *Fish and Fisheries* 17 (4): 1216–1226.

Caddy J.F., and R. Mahon. 1996. *Puntos de referencia para la ordenación pesquera*. FAO documento técnico de pesca 347, 109 p.

Clark, C.W. 1990. *Mathematical bioeconomics: the optimal management of renewable resources (Hard-cover)*. 2nd ed. New York, NY: Wiley, 400 p.

FAO. 1995. *Code of conduct for responsible fisheries*. Rome, Italy: Food & Agriculture Organization.

Froese, R., C. Walters, D. Pauly, H. Winker, O.L.F. Weyl, N. Demirel, A. Tsikliras, and S. Holt. 2016. A critique of the balanced harvesting approach to fishing. *ICES Journal of Marine Science* 73 (6): 1640–1650.

Gabriel, W.L., and P.M. Mace. 1999. A review of biological reference points in the context of the precautionary approach. In *Proceedings of the fifth national NMFS stock assessment workshop: providing scientific advice to implement the precautionary approach under the Magnuson-Stevens fishery conservation and management act*, 34–45. NOAA Tech Memo NMFS-F/SPO-40.

Garcia, S.M., J. Kolding, J. Rice, M.-J. Rochet, S. Zhou, T. Arimoto, J.E. Beyer, L. Borges, A. Bundy, D. Dunn, E.A. Fulton, M. Hall, M. Heino, R. Law, M. Makino, A.D. Rijnsdorp, F. Simard, and A.D.M. Smith. 2012. Reconsidering the consequences of selective fisheries. *Science* 335 (6072): 1045–1047.

Gulland, J., and L.K. Boerema. 1973. Scientific advice on catch levels. *Fishery Bulletin* 71 (2): 325–335.

Hilborn, R. 2007. Reinterpreting the state of fisheries and their management. *Ecosystems* 10 (8): 1362–1369.

Jacobsen, N.S., H. Gislason, and K.H. Andersen. 2013. The consequences of balanced harvesting of fish communities. *Proceedings of the Royal Society B: Biological Sciences* 281 (1775): 20132701.

Katsukawa, T. 2005. Evaluation of current and alternative fisheries management scenarios based on spawning-per-recruit (SPR), revenue-per-recruit (RPR), and yield-per-recruit (YPR) diagrams. *ICES Journal of Marine Science* 62 (5): 841–846.

Mardle, S., and S. Pascoe. 1999. A review of applications of multiple-criteria decision-making techniques to fisheries. *Marine Resource Economics* 14 (1): 41–63.

Munro, G.R. 2000. The United Nations Fish Stocks Agreement of 1995: history and problems of implementation. *Marine Resource Economics* 15 (4): 265–280.

Schaefer, M.B. 1954. Some aspects of the dynamics of populations important to the management of commercial marine fisheries. *Bulletin of Inter-American Tropical Tuna Commission* 1: 27–56.

Ulanowicz, R. 1986. *Growth and development: ecosystems phenomenology*. New York, NY: Springer, 203 p.

———. 2009. The dual nature of ecosystem dynamics. *Ecological Modelling* 220: 1886–1892.

Ulanowicz, R.E. 2018. Biodiversity, functional redundancy and system stability: subtle connections. *Journal of the Royal Society Interface* 15 (147): 20180367.

Chapter 8
Managing Fisheries Under a Holistic Approach

The Challenge of Ecosystem-Based Fisheries Management

Currently, the ecosystem knowledge-based fisheries management tends to have several connotations, i.e., from the target populations to the society, which uses these resources in different ways. In the context of fisheries, one of the major concerns is the negative impacts of fishing on ecosystems, not only overfishing or the overexploitation, but also due to negative effects on other nontarget species, many of them with no commercial value but of importance in other aspects; species of interest in terms of biodiversity and conservation; threatened species, or species that play important roles in an ecosystem; as well as unwanted effects on habitats (Larkin 1996; Schneiders et al. 2012). Considering only the natural system, that is, individuals, populations, and marine ecosystems, the usual ecosystem-based fisheries management approaches focus on some few basic aspects: the target and associated species, as well as the habitat.

Although management necessarily implies the dynamic sustainability of the use of ecosystem resources, it is essentially oriented towards social and economic benefits, whether in terms of social welfare, food provisions, or as a business. Thus, although there may be different goals, in the sense of conventional theoretical modeling, it has generally been assumed, as reference, the maximization, either of returns in terms of volume or economic or in the generation of wealth or jobs; and even, the multiple-optimization. However, the essence in all cases is to maintain sustainable fishing within the context of the production capacity of species and their recovery and, from there, to achieve the intended objectives. In other words, the goal of management is not only the preservation of natural resources; it must consider how to take better advantage of the natural production capacity to achieve desired outcomes; that is, management is for society.

Regarding target populations and from an ecosystem perspective, it is not only overfishing that is of concern, but also overexploitation that could affect the organization and function of ecosystems. For example, the role of large predators in

F. Arreguín-Sánchez, *Holistic Approach to Ecosystem-Based Fisheries Management*, https://doi.org/10.1007/978-3-030-96847-2_8

maintaining ecosystem organization, known as top-down control has been argued, illustrating the negative effect called "fishing down the food web" (Cury et al. 2003; Arreguín-Sánchez 2011), where low trophic species tend to predominate due to the disappearance of top predators. The effects are reflected by the decreasing trends for the average trophic level of the catch and the ecosystem (Pauly et al. 1998; Pauly and Palomares 2005; Liang and Pauly 2020) and in the explosive abundance of lower-trophic level populations, such as jellyfish (Lynam et al. 2006; Jiang et al. 2008). In other cases, concern has been expressed about the intense exploitation of species at the bottom of the trophic pyramid, as in the case of small pelagic fishes, for which the concern has focused on identifying how much biomass should remain in the sea, after fishing, to maintain the organization and function of ecosystems (Roux et al. 2013; Arreguín-Sánchez et al. 2021). In both the cases, however, the management strategies could be oriented to the identification of a limit harvest rate that allows maintaining a biomass in the sea that in turn allows the sustainability of the ecosystem and the resources.

For nontarget species associated with catches, the aim is usually to reduce their retention as much as possible, that is, to reduce bycatch and discards as much as possible. This has two important implications, one associated with the fact that some bycatches are target species for other fisheries, such as the snappers in the northern Gulf of Mexico where overfishing has been reported (Cowan et al. 2011; Pace 2009), and other species that are caught during shrimp trawling and are also targets of small-scale fishing (Shester and Micheli 2011; Eighani and Paighambari 2013; Jones et al. 2018). In such cases, management measures aimed at restoring populations should consider the effects of both fleets. Other management strategies can be aimed to the protection of endangered, threatened, or protected (ETP) species, such as sea turtles, marine mammals, sharks, and billfishes (Watson et al. 2005; Cooke and Ranere 1999; Miller et al. 2016). The bycatch from fish and shrimp trawling fisheries, for example, are of particular concern as their bycatch contains a large amount of biomass of nontarget species (Lewison et al. 2004; Diamond 2004; Kennelly 1995; Kennelly and Broadhurst 2021; Peckham et al. 2007; Alava et al. 2019). Similarly, in many small-scale and industrial fisheries, a series of measures have been taken to protect habitats, with essential measures applied to breeding and reproduction spaces, and to species where habitat disturbance directly affects their population survival (Shester 2008; Shester and Micheli 2011; Hamel et al. 2013). An important aspect of management, in this context, is that in the vast majority of fisheries in the world, there is bycatch, and this occurs essentially because fishing is a probabilistic event and because no species exists in isolation. What is important to consider, in contrast, is that specific actions can be taken as part of the management strategies that minimize bycatch, making fishing an activity as least disturbing as possible for ecosystems.

The common factor of the ecosystem-based management approaches, in the above illustrated cases, is that management measures are typically geared towards the target species and, in the best of cases, to the species associated with the target species or those species that are vulnerable to fishing gears; however, they do not consider the entire ecosystem, itself, as a functional unit in nature. The conventional

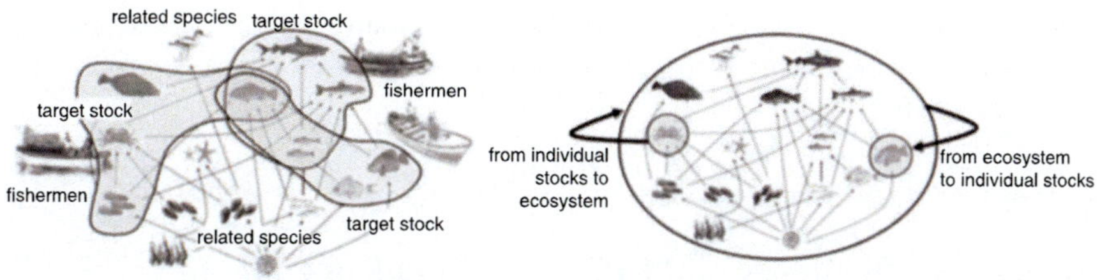

Fig. 8.1 Transiting from a conventional ecosystem approach to fisheries management based on the effect of fishing on components of the ecosystem, to a holistic ecosystem-based management based on trophic network indicators that exhibit impacts of fishing on organization and function of the ecosystem. The former implies definition of harvest rules depending on the impact of fishing; that is, from the individual populations to the ecosystem. The last implies definition of harvest reference points, limit or target, from the holistic properties of the ecosystem to specific stocks

approach can, of course, define mortality controls for populations but does not incorporate any attribute or aspect related to the organization and function of the ecosystem. In this sense, there are key questions that remain unanswered; for example, how much biomass must remain in the sea, after fishing, so as not to affect the organization and function of the ecosystem? The answer to this question, for some species at low trophic levels (small pelagics and shrimp) or top predators (sharks and grouper), can be crucial in terms of maintaining the sustainability of an ecosystem. The condition described above when management is essentially focused on the target species, or to the bycatch, is defined here as a management approach "from the individual populations to the ecosystem"; and this is not properly an ecosystem approach, since it does not consider holistic ecosystem attributes to define management actions on the target species; that is, it does not provide a management approach "from the ecosystem to the individual populations" (Fig. 8.1).

Climate Effects Assumed Overfishing

An additional problem with the "from the individual populations to the ecosystem" approach is that it is typically assumed that changes in population abundance trends are due to fishing, ignoring the effects inherent to the environment and the role of species in an ecosystem. The general problem with this assumption is that it is assumed that management measures will allow the restoration of populations to previous levels of abundance. When stability is assumed and there are medium- or long-term changes in trends for exploited populations and the cause of the trend is fishing, then management measures aimed at restoring populations should work. However, if part of what causes this trend is not due to fishing, such as changes due to climate patterns, management measures will not work according to the expected restoration goals and, in some cases, not even a little. In this scenario, what is relevant is gaining knowledge that allows separating the sources of variation to prevent, with sufficient confidence, the real effects of fishing and to define a strategy of adaptability to the new natural condition of the resource. In this context, fishing controls will not be expected to restore the population to its initial state, just that maintain sustainability.

Overfishing Diagnosis and Effect of Change in Climate Patterns: The Northern Gulf of California

An example of this type of situation is observed from the data reported by Sala et al. (2004), where for the northernmost region of the Gulf of California, the relative species composition of top predators are compared for four consecutive decades through information obtained from interviews with fishermen. These authors report changes in the relative abundance of species between the 1970s and 1980s compared to the 1990s and 2000s (Fig. 8.2a), showing evident changes in the structure of the communities from one period to another. According to the authors, excessive and poorly controlled fishing caused the overfishing of several species during the first two decades, which, by losing their restoration capacity, led to a significant change in the structure of the communities. Under this premise, the authors argue the need for greater fishing control, even considering the possible recommendation of closing some fisheries to restore the current communities to the structure that was had in previous decades.

Subsequently, Arreguín-Sánchez et al. (2017) suggested that these changes in the structure of communities could have been due, largely, to changes in the ecosystem organization influenced in turn by changes in climate patterns, arguing, based on sea surface temperature anomalies, that the presence of the two periods identified corresponds to two different climate regimes, as observed in Fig. 8.2. The alternative hypothesis to that proposed by Sala et al. (2004) is that the change in climate pattern influenced populations and that these populations responded by adapting, reconfiguring the ecosystem. However, the effects of fishing should not be ignored; through changes in abundance patterns due to climate, fishery management, at that time, did not have the capacity to implement adaptability strategies, and assuming that some resources were fully exploited, those whose abundance strongly decreased were probably overfished. Although the two approaches coincide in that some stocks could have been overfished, the difference is in the expected management outcomes. In the first case, when the environment is ignored, resources will never be restored to the previous conditions of decades before, which can be interpreted as a failure in decision-making, and in restrictive measures for fishermen that impact on fisher's communities. If the approach instead considers the environment, probable changes in abundance and exploitation schemes can be defined under an adaptability regime.

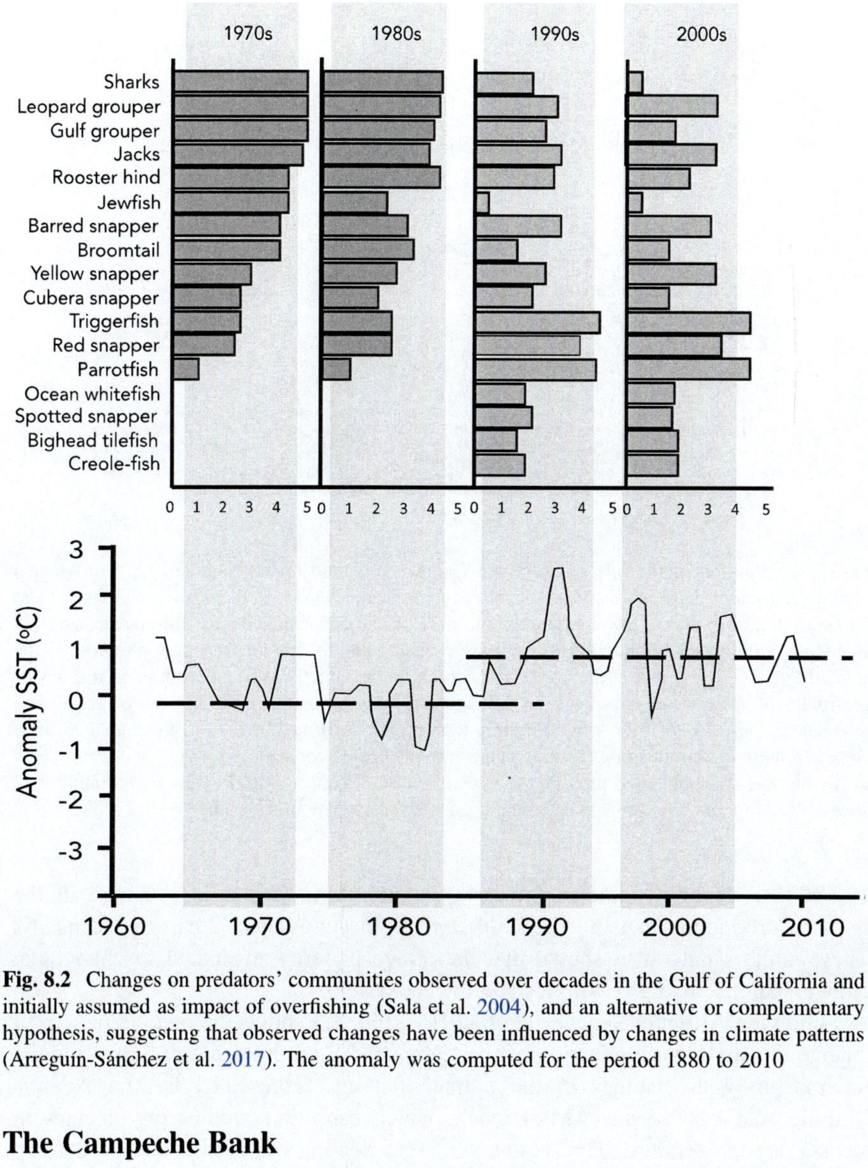

Fig. 8.2 Changes on predators' communities observed over decades in the Gulf of California and initially assumed as impact of overfishing (Sala et al. 2004), and an alternative or complementary hypothesis, suggesting that observed changes have been influenced by changes in climate patterns (Arreguín-Sánchez et al. 2017). The anomaly was computed for the period 1880 to 2010

The Campeche Bank

Another example are the changes observed in the Campeche Bank ecosystem in the southern region of the Gulf of Mexico. By 1995, the collapse of the pink shrimp fishery (*Farfantepenaeus duorarum*) was formally recognized. From 1950s to early 1970s, annual yields were about 18,000 tons per year, declining after that, to less than 2000 tons in 1995 and maintaining such level up to date. Arreguín-Sánchez et al. (2008, 2015, 2019) demonstrated that the collapse of the population was primarily due to a decreasing trend in the recruitment rate, which began in the

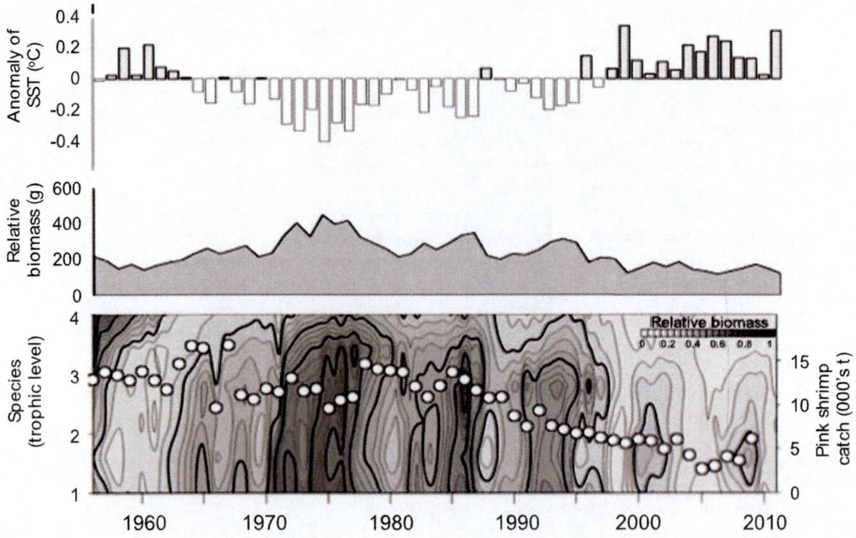

Fig. 8.3 Evolution of the ecosystem of the Campeche Sound on the southern Gulf of Mexico during six decades. Different grays in the low panel indicate changes in the relative biomass, where the clearest corresponds to lower amount of biomass. White dots indicate pink shrimp catches. Gray area in the middle panel represents the carrying capacity, and the bar diagram on top represents the anomaly of SST. Note that at the end of the cooling period, the ecosystem production reaches its maximum production, decreasing later following the warming tendency. Also note the decadal fluctuations but always following the large-term trend. Shrimp population responded to such climate pattern by diminishing its abundance and catches as consequence. The production panel along the time was simulated through the Ecosim model (Walters et al. 1997), incorporating time series of fishing and environmental forcing. (Adapted from Arreguín-Sánchez et al. 2015)

mid-1970s; strongly associated with a medium-term decreasing change in the primary production causing the reduction of the ecosystem's carrying capacity. These authors have also shown that the above patterns are linked to changes in climate patterns acting on species and the ecosystem.

Following Arreguín-Sánchez et al. (2015), Fig. 8.3 shows the relative pattern of change in ecosystem production of the Campeche Bank in recent decades, and the relationship with changes in the climate pattern, represented by the Atlantic Multidecadal Oscillation (AMO) index, which essentially represents changes in sea surface temperature. The regime shift representing the change from a cooling period to a warming period, occurred in the late-1970s; when the highest ecosystem production in the last six decades was also observed; and decreasing towards recent years, showing in addition an evident decadal fluctuation. The outstanding aspect in terms of fisheries management is that since the mid-1990s, with the declaration of the collapse of the pink shrimp fishery, it was assumed that the decline in the population was associated with fishing, and specific measures were taken to restore the stock abundance. Currently, these measures, which are still in force, aim to recover the yields per fishing-day to that observed in the second half of the 1990s (DOF 2012).

These measures include a sharp decrease in the number of trawling boats. The central problem of this fishery is that after almost 25 years, the state of the resource remains the same, despite having met the established management measures, which include avoiding overfishing, both in growth and recruitment, the strict limitation of fishing mortality and the protection of breeding areas. In this case, as demonstrated later by Arreguín-Sánchez (2010) and Arreguín-Sánchez et al. (2015), the decline in the stock was due to the influence of climate, leading to ineffective management measures in terms of the expected outcome, which was to increase the abundance of the shrimp stock. Currently, there is other evidence of the effect of climate on other species, e.g., hawksbill sea turtle (del Monte-Luna et al. 2012), red grouper (Arreguín-Sánchez et al., 2019), red octopus (Arreguín-Sánchez 2019), and spiny lobster (Arreguín-Sánchez et al. 2019), among others. What manifests through this evidence is that the ecosystem has gradually been reconfigured, apparently induced by the climate; and, management measures should reconsider this evidence together with the effects of fishing, to be able to exploit fish resources sustainably under adaptability strategies, and accepting that ecosystem is under a reconfiguration process.

To date, the management objective for pink shrimp fishery continues to restore the stock to the abundance observed in the early 1990s by applying measures such as maintaining reproductive viability and reducing fishing mortality. The pink shrimp stock, to date, remains the same after more than 25 years, despite that the management measures have been applied and respected and that it is a *r*-strategist species, with high reproductive potential and fast individual growth. In this context, there are several aspects related to the effect of climate and the ecosystem to consider. In the first case, Arreguín-Sánchez (2012, 2014, 2019) shows the effect of changes in climate patterns on the ecosystem based on different variables (Fig. 8.3) and relate them to changes in the catch trends for several populations with different life histories, such as mentioned above for the pink shrimp, *Farfantepenaeus duorarum*, red octopus, *Octopus maya*, red grouper, *Epinephelus morio*, spiny lobster, *Panulirus argus*, and also nestings of hawksbill sea turtle, *Eretmochelys imbricata*. Additionally, Del Monte-Luna et al. (2015) show that the harmonic component with a period of 67 years of sea surface temperatures in the Bank of Campeche explains just over 60% of the total variation in sea surface temperature in this region. The contribution of changes in abundance because of environmental effects are not considered in the decision-making process, such that the bias and uncertainty regarding the effectiveness of management measures in meeting expectations is unknown. Moreover, the long-term trend of climate patterns suggests that populations respond in the short term to climate effects, and consequently, ecosystems continuously reconfigure. From this perspective, the scientific basis of management should consider the harvest rate (HR = catch/biomass) as reference, since it can be identified the biomass that must remain in the sea to not affect the functioning and organization of an ecosystem. Since HR is a proportion, it is independent of the magnitude of the abundance and availability itself; and can be adapted to the observed biomass variations. Of course, the harvest rate should be defined according to the life history of the target species, their productive capacity and their role in the ecosystem. For this type of strategy, once the appropriate harvest rate has been identified, the available biomass of the target resources should be estimated each fishing season, and the best way to obtain that biomass in terms of specific tactics based on population attributes should be determined (e.g., to avoid recruitment and growth overfishing). Doing so will allow a dynamic sustainability. Of course, not all ecosystems are equally vulnerable, and management strategies based on holistic ecosystem attributes should consider this fact. The vulnerability of an ecosystem is defined by its sensitivity to stresses, disturbances, or harm and is directly related to the effect of such disturbances on the organization and functioning of the ecosystem. In this context, Ulanowicz (2009) proposes that the aptitude or state of an ecosystem depends on its degree of organization and, as defined in Chap. 3 (Eq. 3.4), reflects the capacity of the ecosystem to change or self-organize. This capacity for self-organization is, in turn, closely related to resilience.

Ecosystem Vulnerability

To a large extent, resilience, as an attribute of an ecosystem, depends on the energy available for this purpose, referred by Ulanowicz (2009) as overhead, and varies according to the distribution of energy flows in the ecosystem. This author suggests that highly organized systems have a strong limitation of flows, characterized by the presence of specialist species from the trophic point of view, leaving little energy available, in the form of conditioned entropy, to contend with disturbances. This property reflects a system with a high degree of vulnerability to disturbances. A typical example in marine ecosystems are coral reefs. In contrast, a system with relatively little organization uses a lower proportion of energy to maintain the organization of the ecosystem, leaving much energy, in the form of entropy, which can be used to contend with disturbances, such that the ecosystem turns out to be less vulnerable in terms of maintaining its organization and functioning. Examples of this type of ecosystem are upwelling systems or coastal lagoons (compared with coral reefs), both with surplus primary production, i.e., energy that is not fully used by the ecosystem itself, being typically exporters of production to neighboring ecosystems.

The importance of identifying the relative vulnerability of ecosystems lies in how much a disturbance will affect the ecosystem beyond its resilience threshold. Of course, if the resilience capacity of an ecosystem is low, the risk of being affected is greater under the same level of disturbance than another system with greater resilience. Arreguín-Sánchez and Ruiz-Barreiro (2014) estimated the relative vulnerability for various types of ecosystems represented by trophic models. This estimation is based on simulation experiments of the impact of biomass extraction on functional groups or species regarding how the ecosystem is organized and represents the effect of the relative entropy gain of the system regarding the organizational change causing for a perturbation in species through the following relationship: Arreguín-Sánchez and Ruiz-Barreiro (2014) estimated the relative vulnerability for various types of ecosystems represented by trophic models. This estimation is based on simulation experiments of the impact of the species' loss of biomass and how the ecosystem responds though the measuring the increase in entropy and change in organization as follows:

$$V_{i,\mathrm{rel}} = \frac{(A/C)_{i,\mathrm{sim}} - [< (A/C)_{\mathrm{ini}}]}{(A/C)_{i,\max} - [< (A/C)_{i,\min}]} \tag{8.1}$$

where

$$V_{i,\mathrm{rel}} = \frac{\text{entropy gained(decomposition)}}{\text{magnitude of organizational change}}$$

where $\frac{A}{C}$ is a measure of the organization of the ecosystem (Ulanowicz (1986, 2009), A is ascendency, and C is development capacity. The numerator represents, by difference, the entropy gain when functional group i is disturbed, where $\left(\frac{A}{C}\right)_{i,\mathrm{sim}}$ represents the organization of the disturbed system (in the simulation), while $\left(\frac{A}{C}\right)_{\mathrm{ini}}$ represents the initial organization of the ecosystem, without disturbance. The denominator represents the total magnitude of the organization change of the ecosystem due to the disturbance, given by the difference between the absolute maximum and minimum values of the degree of organization during simulation. The total relative vulnerability of the system, considering all functional groups can be expressed as

$$V_{\mathrm{rel}} = \sum_{i}^{n} \left(\frac{(A/C)_{i,\mathrm{sim}} - [< (A/C)_{\mathrm{ini}}]}{(A/C)_{i,\max} - [< (A/C)_{i,\min}]} \right) \tag{8.2}$$

In this notation, the symbol "<" refers to the use, for vulnerability estimates, of only those values that cause an increase in entropy, ignoring those cases or situations in which entropy decreases; showing with it an essentially precautionary perspective. When this estimate is referred to as a pattern regarding the trophic levels of the different species, the relative contribution of each functional group to the vulnerability is obtained as expressed by Eq. (8.1). Figure 8.4 shows the changes in the relative contribution of the functional groups to the vulnerability of the ecosystem, where it can be observed, by the slopes, the relative differences in the vulnerability between different type of ecosystems, according to their degree of organization.

In Fig. 8.4, two aspects are highlighted: systems are generally more vulnerable when disturbances originate at low trophic levels, and ecosystems with greater organization, such as coral reefs, are more vulnerable than less organized systems, such as coastal lagoons, and continental shelves exhibit intermediate vulnerability. In this context, Arreguín-Sánchez and Ruiz-Barreiro (2014) suggest that vulnerability is proportionally related with the organization and complexity of the ecosystems. This is initially supported by the relationship with some ecosystem attributes, such as

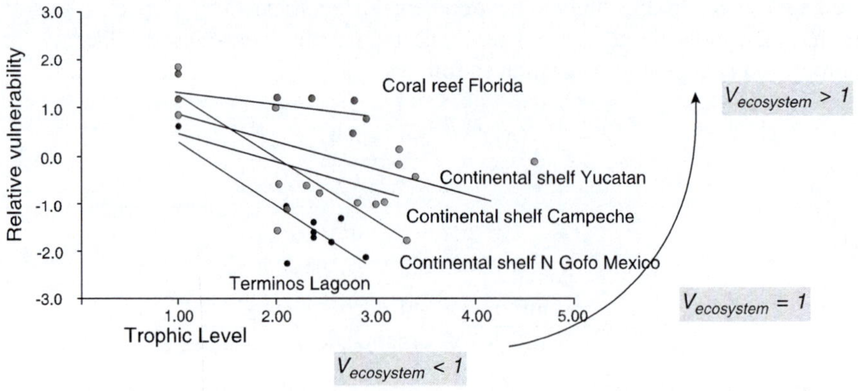

Fig. 8.4 Relative vulnerability for several marine ecosystem. Slopes of lines of tendencies represent the ecosystem vulnerability. Note that complex ecosystems, represented in the picture by a coral reef model are highly vulnerable (vulnerability index >1) than the less complex (vulnerability index <1), exemplified with a coastal lagoon. Mid vulnerable ecosystems are all exemplified by continental shelves. The five ecosystems illustrated are from the Gulf of Mexico (for more details, see Arreguín-Sánchez and Ruiz-Barreiro 2014)

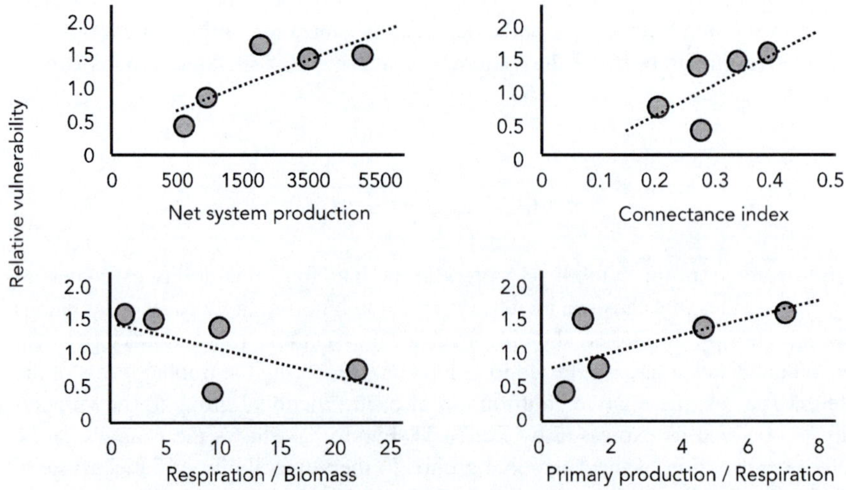

Fig. 8.5 Relationships among the relative ecosystem's vulnerability and functional ecosystem attributes. (Adapted from Arreguín-Sánchez and Ruiz-Barreiro 2014)

connectivity, efficiency, and net production; and inversely with the energy expenditure, represented by the respiration/biomass ratio (Fig. 8.5).

Likewise, in Fig. 8.4, it can be observed that the gap in ecosystem vulnerability when different trophic levels are disturbed is greater in more organized ecosystems than in less organized ecosystems. This implies that in ecosystems such as coral reefs, which are more organized, the vulnerability of the ecosystem tends to be high

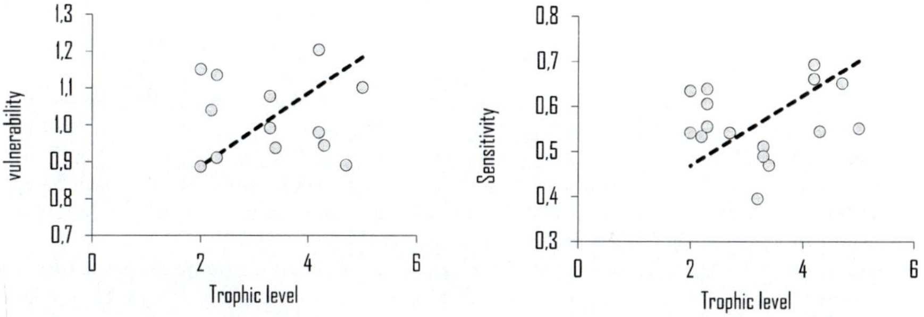

Fig. 8.6 Relationship between vulnerability and sensitivity of species with trophic level. Note that both the attributes increase with trophic level a rate about 10%

and similar for the different trophic levels; this is not observed in ecosystems with less organization. This behavior is strongly associated, in its origin, with the conditional flows and the consequent lower amount of overhead (conditional entropy) available in the ecosystem; in other words, its resilient capacity. Also, for all the cases, ecosystems are more vulnerable to perturbations on low trophic levels than higher ones.

The vulnerability patterns observed in Figs. 8.4 and 8.5 allow us to understand the importance of considering the limit ecosystem-based biological reference point (EBRP-limit) as a management criterion in fisheries, given that the resilience threshold of the different functional groups of an ecosystem is not similar and, consequently, the potential effect of a disturbance will depend on the species or functional group affected. Thus, for example, the high resilience threshold of a low trophic level contrasts with the relatively low vulnerability of the ecosystem, with the opposite occurring with high trophic levels. This characteristic corresponds to the particular sensitivity of trophic levels, as shown in Fig. 8.6. Such behavior patterns have great importance, from the management perspective, in responses to scenarios such as climate change and adaptability through strategies and tactics to be applied; and in particular, their understanding serves as the basis for preventative management strategies.

In conclusion, understanding the dynamics of ecosystems and the behavior patterns of their essential attributes of organization and function are key to the use of ecosystem-based reference points, mainly under precautionary management strategies. This, in addition to the elements conventionally applied at the population level. It is important to consider that current management strategies do not consider EBRPs; however, these are considered high priority for defining fishing strategies that ensure that the effects of biomass extraction do not affect, after fishing, the organization and functioning of ecosystems. The effects of short- and medium-term natural climate variation patterns, as well as the effects of climate change, cause a continuous process of reconfiguration of ecosystems; in this way, the explicit recognition of EBRPs under an adaptability strategy acquires greater relevance for sustainability, in contrast to the passive management conventionally applied.

References

Alava, J.J., B. Tatar, M.J. Barragán, C. Castro, P. Rosero, J. Denkinger, et al. 2019. Mitigating cetacean bycatch in coastal Ecuador: governance challenges for small-scale fisheries. *Marine Policy* 110: 102769.

Arreguín-Sánchez F., M. Zetina-Rejón, M. Ramírez-Rodríguez. 2008. Exploring ecosystem-based harvesting strategies to recover the collapsed pink shrimp (Farfantepenaeus duorarum) fishery in the southern Gulf of Mexico. Ecological Modelling 214(2–4): 83–94.

Arreguín-Sánchez, F. 2010. *Cambio climático y el colpaso de la pesquería de camarón rosado (Farfantepenaeus duorarum) de la sonda de Campeche*. Cambio Climático en México un Enfoque Costero-Marino. Universidad Autónoma de Campeche, 399–410.

———. 2011. Ecosystem dynamics under "top-down" and "bottom-up" control situations generated by intensive harvesting rates. *Hidrobiológica* 21 (3): 323–332.

———. 2012. The dynamics linking biological hierarchies, fish stocks and ecosystems: implications for fisheries management. In *Models of the ecological hierarchy: from molecules to the ecosphere*, ed. F. Jordán and S.E. Jørgensen, 501–516. Oxford, UK: Elsevier B.V.

———. 2014. Adaptabilidad y sostenibilidad: un enfoque holístico hacia el manejo de ecosistemas acuáticos explotados. *Ciencia Pesquera* 22 (1): 89–101.

———. 2019. Climate change and the rise of the octopus fishery in the Campeche Bank, Mexico. *Regional Studies in Marine Science* 32: 100852.

Arreguín-Sánchez, F., and T.M. Ruiz-Barreiro. 2014. Approaching a functional measure of vulnerability in marine ecosystems. *Ecological Indicators* 45: 130–138.

Arreguín-Sánchez, F., P. del Monte-Luna, and M.J. Zetina-Rejón. 2015. Climate change effects on aquatic ecosystems and the challenge for fishery management: pink shrimp of the Southern Gulf of Mexico. *Fisheries* 40 (1): 15–19.

Arreguín-Sánchez, F., P. del Monte-Luna, M.J. Zetina-Rejón, and M.O. Albáñez-Lucero. 2017. The Gulf of California large marine ecosystem: fisheries and other natural resources. *Environmental Development* 22: 71–77.

Arreguín-Sánchez, F., M.O. Albañez-Lucero, P. Del Monte-Luna, and M.-J. Zetina-Rejón. 2019. Fishery resource management challenges facing climate change. In *Mexican aquatic environments*, ed. A.L. Ibañez, 181–194. New York, NY: Springer.

Arreguín-Sánchez, F., P. del Monte-Luna, M.O. Albañez-Lucero, M.J. Zetina-Rejón, A. Tripp-Quezada, T.M. Ruiz-Barreiro, and J.C. Hernández-Padilla. 2021. How much biomass must remain in the sea after fishing to preserve ecosystem functioning? The Case of the Sardine Fishery in the Gulf of California, Mexico. In *Marine coastal ecosystems modelling and conservation*, ed. M. Ortiz and F. Jordán, 143–161pp. Cham, Switzerland: Springer.

Cooke, R.G. and A.J. Ranere 1999. Precolumbian fishing on the Pacific coast of Panama. 103–121pp. In: M. Blake (ed.). *Pacific Latin America in Prehistory: the evolution of archaic and formative cultures*. Washington State University Press, Pullman, Washington.

Cowan, J.H., C.B. Grimes, W.F. Patterson, C.J. Walters, A.C. Jones, W.J. Lindberg, D.J. Sheehy, W.E. PineIII, J.E. Powers, M.D. Campbell, K.C. Lindeman, S.L. Diamond, R. Hilborn, H.T. Gibson, and K.A. Rose. 2011. Red snapper management in the Gulf of Mexico: science- or faith-based? *Reviews in Fish Biology and Fisheries* 21 (2): 187–204.

Cury P., L. Shannon, Y.J. Shin, M. Sinclair, M.G. Valdimarsson. 2003. The functioning of marine ecosystems: a fisheries perspective. *Responsible fisheries in the marine ecosystem*, Cambridge, MACABI Pubkishing, 426p

Diamond, S.L. 2004. Bycatch quotas in the Gulf of Mexico shrimp trawl fishery: can they work? *Reviews in Fish Biology and Fisheries* 14 (2): 207–237.

DOF. 2012. *Carta Nacional Pesquera*. Mexico: Diario official de la Federación, 24 August 2012. www.conapesca.sagarpa.gob.mx.

Eighani, M., and S.Y. Paighambari. 2013. Shrimp, bycatch and discard composition of fish caught by small-scale shrimp trawlers in the Hormuzgan Coast of Iran in the Persian Gulf. *The Philippine Agricultural Scientist* 96 (3): 314–319.

Hamel, M.A., S. Andréfouët, and R.L. Pressey. 2013. Compromises between international habitat conservation guidelines and small-scale fisheries in Pacific island countries. *Conservation Letters* 6 (1): 46–57.

Jiang, H., H.Q. Cheng, H.G. Xu, F. Arreguín-Sánchez, M.J. Zetina-Rejón, P.D.M. Luna, and W.J. Le Quesne. 2008. Trophic controls of jellyfish blooms and links with fisheries in the East China Sea. *Ecological Modelling* 212 (3-4): 492–503.

Jones, B.L., R.K. Unsworth, S. Udagedara, and L.C. Cullen-Unsworth. 2018. Conservation concerns of small-scale fisheries: by-catch impacts of a shrimp and finfish fishery in a Sri Lankan lagoon. *Frontiers in Marine Science* 5: 52.

Kennelly, S.J. 1995. The issue of bycatch in Australia's demersal trawl fisheries. *Reviews in Fish Biology and Fisheries* 5 (2): 213–234.

Kennelly, S.J., and M.K. Broadhurst. 2021. A review of bycatch reduction in demersal fish trawls. *Reviews in Fish Biology and Fisheries* 31: 1–30.

Larkin, P.A. 1996. Concepts and issues in marine ecosystem management. *Reviews in Fish Biology and Fisheries* 6 (2): 139–164.

Lewison, R.L., L.B. Crowder, A.J. Read, and S.A. Freeman. 2004. Understanding impacts of fisheries bycatch on marine megafauna. *Trends in Ecology & Evolution* 19 (11): 598–604.

Liang, C., and D. Pauly. 2020. Masking and unmasking fishing down effects: the Bohai Sea (China) as a case study. *Ocean & Coastal Management* 184: 105033.

Lynam, C.P., M.J. Gibbons, B.E. Axelsen, C.A. Sparks, J. Coetzee, B.G. Heywood, and A.S. Brierley. 2006. Jellyfish overtake fish in a heavily fished ecosystem. *Current Biology* 16 (13): R492–R493.

Miller, K., R. Jauharee, I. Nadheeh, and M.S. Adam 2016. *Interactions with Endangered, Threatened and Protected (ETP) species in the Maldivian pole-and-line tuna fishery*. IPNLF and MRC.

del Monte-Luna, P., V. Guzmán-Hernández, E.A. Cuevas, F. Arreguín-Sánchez, and D. Lluch-Belda. 2012. Effect of North Atlantic climate variability on hawksbill turtles in the Southern Gulf of Mexico. *Journal of Experimental Marine Biology and Ecology*. 412: 103–109.

del Monte-Luna, P., H. Villalobos, and F. Arreguín-Sánchez. 2015. Variability of sea surface temperature in the southwestern Gulf of Mexico. *Continental Shelf Research* 102: 73–79.

Pace, N.L. 2009. Ecosystem-based management under the Magnuson-Stevens Act: managing the competing interests of the Gulf of Mexico red snapper and shrimp fisheries. *Sea Grant Law & Policy Journal* 2 (2): 1–28.

Pauly, D., and M.L. Palomares. 2005. Fishing down marine food web: it is far more pervasive than we thought. *Bulletin of Marine Science* 76 (2): 197–212.

Pauly, D., V. Christensen, J. Dalsgaard, R. Froese, and F. Torres. 1998. Fishing down marine food webs. *Science* 279 (5352): 860–863.

Peckham, S.H., D.M. Díaz, A. Walli, G. Ruiz, L.B. Crowder, and W.J. Nichols. 2007. Small-scale fisheries bycatch jeopardizes endangered Pacific loggerhead turtles. *PLoS One* 2 (10): e1041.

Roux, J.P., C.D. van der Lingen, M.J. Gibbons, N.E. Moroff, L.J. Shannon, A.D. Smith, and P.M. Cury. 2013. Jellyfication of marine ecosystems as a likely consequence of overfishing small pelagic fishes: lessons from the Benguela. *Bulletin of Marine Science* 89 (1): 249–284.

Sala, E., O. Aburto-Oropeza, M. Reza, G. Paredes, and L.G. López-Lemus. 2004. Fishing down coastal food webs in the Gulf of California. *Fisheries* 29 (3): 19–25.

Schneiders, A., T. Van Daele, W. Van Landuyt, and W. Van Reeth. 2012. Biodiversity and ecosystem services: complementary approaches for ecosystem management? *Ecological Indicators* 21: 123–133.

Shester, G.G. 2008. *Sustainability in small-scale fisheries: an analysis of ecosystem impacts, fishing behavior, and spatial management using participatory research methods*. Stanford, CA: Stanford University.

Shester, G.G., and F. Micheli. 2011. Conservation challenges for small-scale fisheries: bycatch and habitat impacts of traps and gillnets. *Biological Conservation* 144 (5): 1673–1681.

Ulanowicz, R. 1986. *Growth and development: ecosystems phenomenology*, 203. New York, NY: Springer.

———. 2009. The dual nature of ecosystem dynamics. *Ecological Modelling* 220: 1886–1892.

Walters, C., V. Christensen, and D. Pauly. 1997. Structuring dynamic models of exploited ecosystems from trophic mass-balance assessments. *Reviews in Fish Biology and Fisheries* 7 (2): 139–172.

Watson, J.W., S.P. Epperly, A.K. Shah, and D.G. Foster. 2005. Fishing methods to reduce sea turtle mortality associated with pelagic longlines. *Canadian Journal of Fisheries and Aquatic Sciences* 62 (5): 965–981.

Chapter 9
Adaptability as Management Strategy

The Need of Adaptability as Management Strategy

One of the inherent characteristics of fishing is that there are no inputs to production, being an essentially harvesting activity, where the yields, expressed as retained biomass, depend on the natural production capacity of species. As mentioned before, during the period from approximately 1930 to 1980, natural systems could be considered statistically stable; and no changes were observed in ecosystems, given the absence of climate change patterns with tendency (see Chap. 2). During this time, fishing was the main driver of the changes in resource abundance, such that depleted populations could be restored by controlling fishing. This notion governed management strategies for several decades, being only necessary to get detailed information on the effects of fishing on the population structure to provide information, increasingly accurate, given the increasing intensity and efficiency of fishing (Fig. 2.1). On the other hand, the natural changes in climate patterns, and the human-induced effects on the climate; defined as climate change by IPCC (Pielke 2004; Solomon et al. 2007); have promoted continuous processes of response, adaptation, and reconfiguration of natural systems. Concerning the management efficiency for living marine resources, the outcome of these processes translates into the need to separately identify both climate-induced and fishing-induced patterns so that available biomass estimation is as precise as possible and that management goals can be adapted, and adequate harvest rates applied.

At the population level, conventional fisheries management tends to be passive; that is, it involves static regulatory measures, such as minimum stock size, closures, or using fixed levels of abundance or fixed levels of fishing effort (e.g., number of boats or fishing permits) and are used together with reference points. These measures, at present, are no longer adequate because in their application, a response from the population is expected that aligns with the assumption that fishing is the only factor that induces significant changes in abundance. What is the consequence of this assumption? Let us imagine that a stock is being exploited to the maximum of its

F. Arreguín-Sánchez, *Holistic Approach to Ecosystem-Based Fisheries Management*, https://doi.org/10.1007/978-3-030-96847-2_9

production capacity, by applying a constant catch, suppose at the beginning that it corresponds to 40% of the available biomass, this happening under a management scheme that defines a stable fishing effort. Let us assume that in a given moment, some environmental driver affect population by diminishing its abundance, with a continuous decreasing trend towards the midterm. If the control measure is not adjusted to the new level of available biomass, the resource will be overfished. Moreover, when overfishing is detected, the assumption, in this hypothetical example, is that this state of the resource, due to overfishing, will require management measures that aim to restore the stock to the state prior to its depleted condition. What is expected of the measures adopted to counteract overfishing is essentially to reduce fishing mortality, with the hope that the population will respond by restoring its initial condition. The usual problem in these cases is that the desired result is not obtained, largely because part of reason for the declining in abundance is due to natural causes and not because of fishing. In this context, the fishing sector tends to spend time waiting for the stock recovery, and this never happens or does so in an unexpected way. Eventually, the user of the resource will initially react by rejecting the management measures, believing they are ineffective, or in the most severe cases, there is distrust in institutions, generating certain levels of ungovernability and even promoting undesirable practices such as noncompliance with norms and illegal activity.

Adaptability and Individual Stocks: The Campeche Bank, Gulf of Mexico

In the following paragraphs, the Campeche Bank region, on the southern Gulf of Mexico, is used to illustrate the problem of fisheries management, especially the need for adaptability strategies, since the effects of the change in climate patterns have clearly manifested in this region, exhibiting a continuous ecosystem reconfiguration over time, and reflecting the challenge that such reconfiguration represents for the sustainable management of fisheries in a dynamic and constantly changing environment. The Bank of Campeche is historically an important fishing region; for decades, it has sustained a dynamic fishing sector. The highest global yields occurred in the decades from 1950 to 1980, subsequently declining significantly until the present time, affecting the main fisheries, such as shrimp, grouper, octopus, lobster, and even species of conservation interest as sea turtles, among others. The changes in the production of the region have been attributed to overfishing, but there is a clear evidence that the observed trends have been strongly induced by changes in climate patterns, especially those associated with a regime shift. Studies at the ecosystem level, supported by population dynamics data, also suggest that the ecosystem has been reconfigured over time, being a clear example of the need for knowledge to guide fishery sustainability via adaptability in the context of ecosystem-based management.

*The Pink Shrimp (*Farfantepenaeus duorarum*) Fishery*

The pink shrimp (*Farfantepenaeus duorarum*) fishery off the Bank of Campeche in the southern Gulf of Mexico (Fig. 9.1) was historically the most important fishery along the Mexican coast, with an average of about 20,000 tons of pink shrimp caught per year from the mid-1950s to the early 1970s, with a fishing effort slightly over 11,000 trips per year. Catches declined gradually, and by the mid-1990s, the collapse of the fishery was declared, with annual catches of just 2500 tons and 1700 fishing trips. Currently, annual yields are just above 1000 tons (Arreguín-Sánchez 2010; Arreguín-Sánchez et al. 2008, 2015, 2017; DOF 2012). In the mid-1990s, the hypothesis of such collapse was overfishing followed by overexploitation, adding to this, the deterioration of the juvenile's nursery areas. The management measures implemented were a reduction in fishing effort, a ban to protect the spawning stock, maintaining a minimum catch size, and prohibiting fishing in nursery areas, at inland and coastal waters within 15 nautical miles. From that time to the present, the stated objective of these management measures is to restore the fishery yields to the abundances seen in the mid-1990s, that is, reach similar catch-per-unit-of-effort (cpue). After 20 years and despite applying and respecting the measures imposed, the relative abundance remains low, just over 5% of the abundance observed in the early 1970s and 30% lower than that presented in the mid-1990 (Fig. 9.1).

From an ecosystem perspective, Fig. 9.2 shows the changes over time in relative biomasses, during the last six decades, associated with changes in climate regime

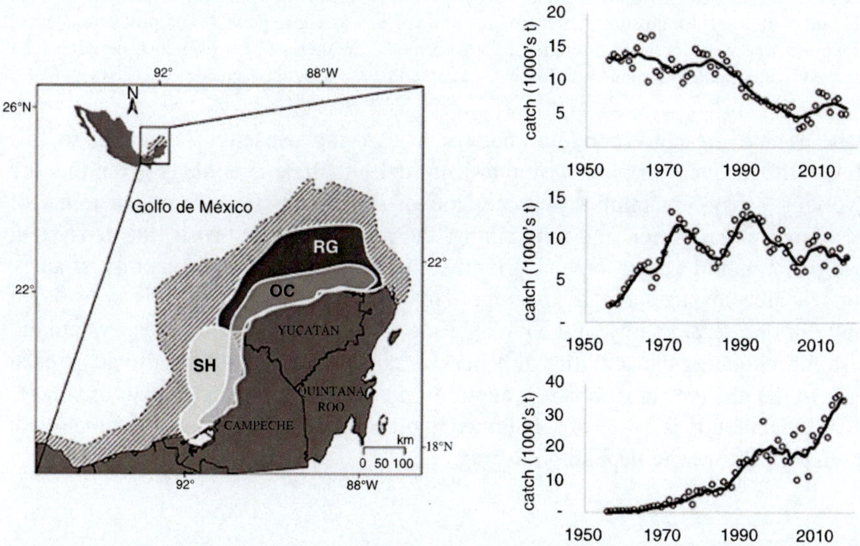

Fig. 9.1 Spatial distribution (left) of the fisheries of red grouper (RG) octopus (OC) and shrimp (SH), the main fish resources in the continental shelf (dashed area) of the Campeche Bank, in the southern Gulf of Mexico. On the right, the historical catch tendencies showing a depletion of shrimp, an increase of octopus and fluctuations of the red grouper with a decrease in last decades

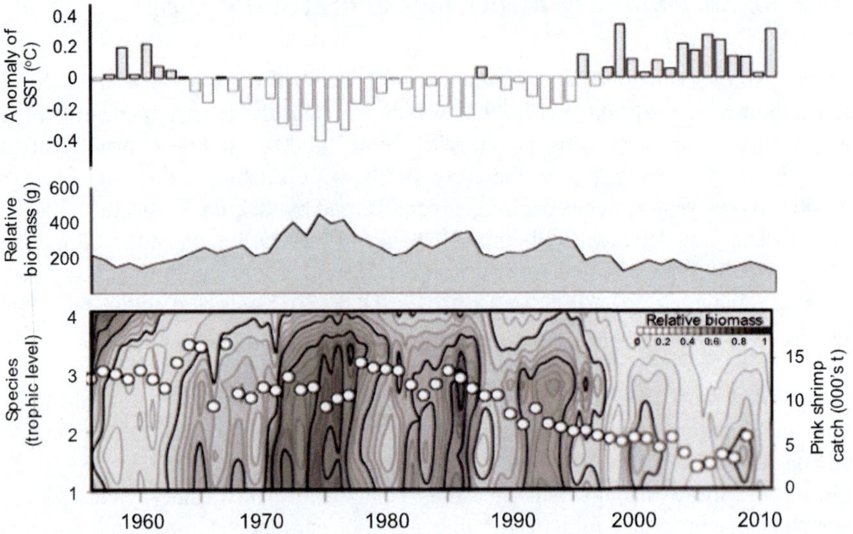

Fig. 9.2 Evolution of the ecosystem of the Campeche Sound on the southern Gulf of Mexico during six decades. Different grays in the low panel indicates changes in the relative biomass, where the clearest corresponds to lower amount of biomass. White dots indicate pink shrimp catches. Gray area on the middle panel represents the carrying capacity, and the bar diagram on top represents the anomaly of SST. Note that at the end of the cooling period, the ecosystem production reaches its maximum production, decreasing later following the warming tendency. Also note the decadal fluctuations but always following the large-term trend. Shrimp population responded to such climate pattern by diminishing its abundance and catches as consequence. The production panel along the time was simulated through the Ecosim model (Walters et al. 1997), incorporating time series of fishing and environmental forcing. (Adapted from Arreguín-Sánchez et al. 2015)

patterns and the corresponding changes in carrying capacity. According to this figure, the recovery of pink shrimp to its mid-1990s state is unlikely given that the evolving ecosystem implies a succession of states that depends on, in terms of ecosystem organization and functioning, the previous state; that is, the ecosystem has reconfigured across time. Currently, the absence of an adaptability strategy for resource management is apparent. This process of adaptability is critical not only for the sustainability of the pink shrimp stock and the exploited ecosystem but also for regulating the activities of fishers, as direct users, who must also adapt each year to the changes in resource availability. In the sense of precautionary management, adaptability is a tool closely linked to planning with direct effects on the social welfare of those who depend on fishing.

*The Red Grouper (*Epinephelus morio*) Fishery*

The grouper fishery (*Epinephelus morio*) (Fig. 9.1) yielded annual catches of approximately 18,000 tons in the mid-1970s and early 1980s, declining to approximately 6000 tons currently (Echazabal-Salazar et al. 2021). This is a multispecific fishery; in its beginnings, *E. morio* represented more than 85% of the catch, while currently, its contribution is approximately 45% of the total catch. As in the previous case of pink shrimp, overfishing was declared in early 1990s. *E. morio* is a protogynous hermaphrodite species that spawns in large aggregations during the winter in the central-eastern region of the continental shelf. It has been reported that the species requires relatively low sea temperatures for reproduction, where spawning aggregations occurs in synchrony with seasonal upwelling in early spring (Arreguín-Sánchez and Pitcher 1999; Brulé et al. 1999; Burgos-Rosas and Pérez-Pérez 2006). The stock assessments reported in the literature (Burgos-Rosas and Pérez-Pérez 2006; Giménez-Hurtado 2005; Echazabal-Salazar et al. 2021) have identified a decrease in population abundance, in both the spawning stock and recruitments, closely following the temperature trend (Fig. 9.3). Three fleets participate in this fishery: a small-scale coastal fleet, a medium-sized fleet and a technified Cuban fleet. It is known that the target stock has been fully exploited since the early 1990s and that the fishing effort has been, in principle, controlled through the number of permits, although the number reported is unknown (DOF 2012). What is known from the species biology indicates that the optimal spawning temperature threshold is approximately 23 °C (Bessonov et al. 1971; González et al. 1974, García 1980), a condition that is unlikely due to the warming trend after the regime shift in 1980. This information, together with stock abundance changes over the last decades, suggests an effect of environmental forcing on the spawning process that

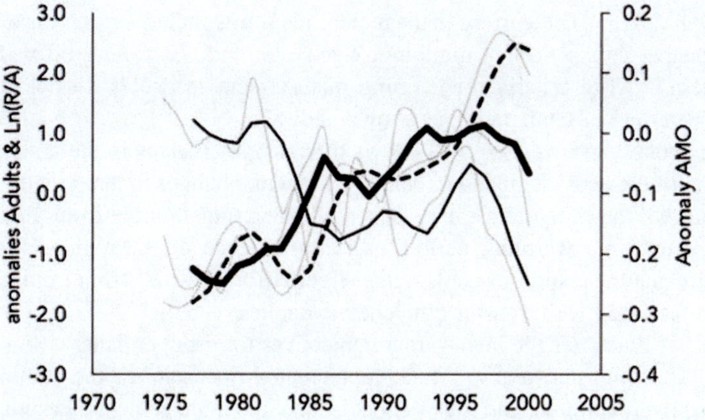

Fig. 9.3 Relationship between the anomaly of SST, given by the AMO (dashed line), with adults (dark bold line) and recruits (thin solid line) of the red grouper of the Campeche Bank, southern Gulf of Mexico. For all the cases, tendencies represent 3 years moving average; dotted thin lines represent observed data

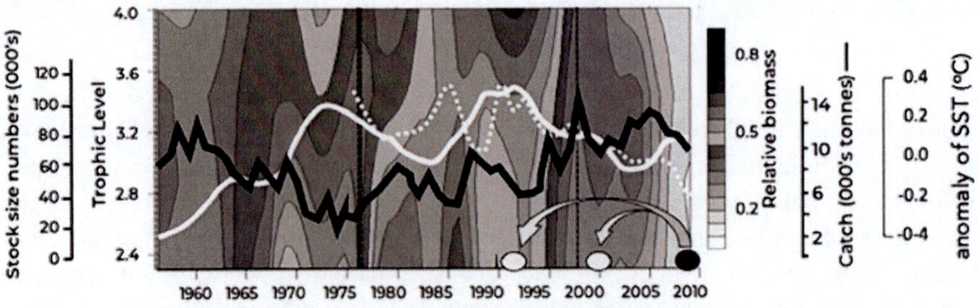

Fig. 9.4 Evolution of the ecosystem of the north continental shelf of Yucatan in the southern Gulf of Mexico during six decades. Different grays indicate changes in the relative biomass, where the clearest corresponds to lower amount of biomass. Black solid line represents the anomaly of the SST; white solid and dotted white lines represent red grouper catches and stock size, respectively. Arrows starting from the recent year to backwards represent the objectives of current management strategies. Here observation is made that the structure of the ecosystem, and with it its function, was different in the times to which it is intended to restore the population of red grouper. This leads to the assumption that the management measures will not have the expected efficiency, since the ecosystem in the near future no necessarily will have the same condition. (Figure was derived from simulation experiments through the Ecosim model (Walters et al. 1997), incorporating time series of fishing and environmental forcing)

is evident in the decreasing recruitment trend, which, sustained over time, adversely impacts stock abundance (Arreguín-Sánchez 2012). This is a phenomenon equivalent to recruitment overfishing and was probably confused with it, given that in practice, until 1994, when the ban was implemented for the first time, fishing on spawning stock occurred annually given the high local yields obtained during the late winter and early spring. Currently, the state of the fishery is considered deteriorated, according to a biomass threshold that was defined based on the estimated maximum sustainable yield corresponding to biomass levels existing in the early 1980s (DOF 2018). The current management measures include spawning closures, quotas for the Cuban fleet, a minimum catch size and controlling effort through fishing permits. The criteria for defining management measures do not take into account the effect of temperature as shown above.

At the ecosystem level, Fig. 9.4 shows the relative changes in the structure and production of the ecosystem and its relationship with changes in temperature, where the response of the ecosystem can be observed when transitioning from the end of a cooling period to a warming period as a consequence of the regime shift. It is noteworthy that the response seems to lag slightly from low to higher trophic levels, possibly associated with transfer efficiency dynamics.

Figure 9.4 illustrates the management objectives for grouper fishery. Since 2003, when the spawning closure was first implemented, the state of the resource was considered overexploited, and it was expected that the management measures would lead to population recovery by reducing fishing mortality and then reducing overfishing, both recruitment overfishing and growth overfishing; however, two decades later, the stock has not been restored and does not seem to have stabilized (Echazabal-Salazar et al. 2021). Figure 9.4 shows that the distribution of biomass by trophic level across time is not stable, and changes in the population, in this case of red grouper, suggest that the ecosystem has been reconfiguring across time and that the aim of restoring the grouper stock to the levels of 1990s, only by controlling fishing mortality, does not seem viable, at least not under the expected action–response criterion. In this case, what seems most congruent is an adaptability strategy for the exploitation of available biomass year after year.

*The Red Octopus (*Octopus maya*) Fishery*

Another case of interest in the same region is the octopus fishery, historically based on the endemic species *Octopus maya*, commonly called the red octopus, which represents approximately 80% of catches, with the remainder corresponding to the common octopus, *O. vulgaris*. In the 1980s, the octopus fishery reported catches of approximately 8000 tons and operated at its maximum production capacity, with the mention of a potential risk of overfishing if relevant measures were not taken (Arreguín-Sánchez et al. 1987). Subsequently, catches have been increasing steadily to between 18,000 and 22,000 tons of this species since the late 1990s to date. In this regard, Arreguín-Sánchez (2019) shows that this trend is significantly associated with the effects of climate pattern changes, particularly during the warming period that began after the regime shift at the beginning of the 1980s, but also with changes in global trend of solar irradiance for the same period, relating both cases with the robustness and growth of *O. maya*. By the 1990s, the resource was considered fully exploited, with recommendations to not increase fishing effort (Arreguín-Sánchez 1987). At that time, management measures consisted of implementing a minimum catch size, closures, and controlling fishing gear. From 2001 to 2003 (DOF 2006), a catch quota was assigned that varied between 11,000 and 13,000 tons. In 2012 (DOF 2012), the same management measures were employed, but a biological reference point was defined as a function of the stock density at 1851 individuals per hectare at the beginning of the fishing season, indicating that the stock is fully exploited. By 2018 (DOF 2018), the resource continued being fully exploited at the maximum sustainable level, defining an annual catch quota per fishing season and maintaining the same management measures. In the last decade, recorded catches exceeded catch quotas, with an average increase rate of 26%/year. In this case, the strategy based on a catch quota per year at the beginning of the fishing season seems to be the most successful in terms of adaptability, given the life history of the species (short-lived), associated with the fluctuating biomass whose availability is adapting, in turn, to the continuous reconfiguration of the ecosystem.

The significant relationship between sea surface temperature, representative of the regime shift, and catch trends of several species in the Banco de Campeche region has been described. According to the trends shown, the regime shift defines a threshold where the trends for population abundance and capture change strongly, reflecting changes in the ecosystem. These changes and ecosystem trends towards the present time establish the context of a continuous ecosystem reconfiguration, to which, fishing must adapt to develop in a sustainable way. In terms of management, it is clear that fishing also contributes to changes in the abundance of populations, making it necessary to dynamically define harvest rates under the premise of identifying the amount of biomass for each resource that should remain in the sea after fishing to maintain ecosystem function and organization; and therefore, a sustainable exploitation. In the context of conventional management, typically passive, management measures such as closure, minimum catch size, or catch quota alone do not contribute to improving management systems because they do not consider the effects of climate patterns, and even less the evolution of ecosystems represented by their continuous reconfiguration. Considering ad hoc management measures of individual populations through adaptability to changing trends is undoubtedly one of the great current challenges associated with the dynamics of sustainability.

Ecosystem's Considerations Towards Adaptability

Another aspect refers to the development of adaptability strategies relating to biological interdependencies or, in other words, to the effects of the ecosystem reconfiguration, particularly because the relationships among species are not linear and the variations in structure and function in different ecosystem states, even when they are successive, are not necessarily similar, thus affecting management decisions. Suppose, for example, that the strategy to be followed is to simultaneously exploit the different species, assuming as target the biological reference point corresponding to the maximum sustainable yield. At this point, it is assumed that the populations maintain their production capacity and replace the biomass loss due to catch per unit of time; even when the populations vary with a trend, if the catch rates are established appropriately and only the adequate proportion of the available biomass is extracted, the forecast is that the sustainability of the populations will be maintained. In this context and respecting the appropriate controls, it is assumed that sustainability in biomass production is effectively maintained.

Although the previous strategy, with the appropriate controls, allows to assume the sustainability of the exploited populations, this is not necessarily the case with the ecosystem because, as seen above, the resilience mechanisms of the ecosystem do not depend only on the populations but on the capacity of the ecosystem itself to maintain its own energy balance and sustainable state. Assuming that the ecosystem maintains its resilience through populations eventually leads to failed expectations regarding how effective management measures are when sustainability is assumed for the ecosystem itself and for fisheries; in other words, controls for sustainability of ecosystem processes through their biological interdependencies and energy balance are relevant, even more so when ecosystems are in a continuous reconfiguration process.

In this way, in the previous condition, stock exploitation is maintained at the maximum biomass production capacity, and the control measures allow the catch to be adapted to this production capacity year after year. Consider now the entropy associated with the resilience capacity of the ecosystem (see Fig. 6.3), represented by the concept of overhead (Ulanowicz 1986, 2009). The overhead is interpreted as the energy in reserve of the ecosystem and is represented by the unconstrained energy flows between components of the ecosystem; this entropy is the resilience capacity of the system. Attempting to maintain the different exploited resources at their maximum production capacity would imply, theoretically, an assumption of ecosystem stability such that the energy balance of the system could sustain a condition where entropy would maintain stable flow probabilities (a contradictory situation by itself). This assumption would imply a new stable state of the ecosystem, where the maximum fishery production would depend on the life history of each species. Thus, there would be resources whose harvest rate could be, for example, HR = 45%, which means that fishing extracts 45% of the available biomass and a certain amount of biomass is replenished per unit of time, while for another species, it could be only HR = 2% or HR = 25%, and so on for each species.

Suppose now that some type of contingency occurs that negatively affects the biomass of some species or an individual species, such that the energy balance in this system is altered. If the species are being exploited at their maximum production capacity and a contingency occurs in these conditions, the remaining production capacity can no longer compensate for the total loss, but only that which its production capacity allows per unit of time. Given this effect, the distribution of the conditional flows of the ecosystem changes as a function of the capacities defined by each independent predator–prey relationship, that is, food preferences defined as differences in the conditionality of flows, and the ability of the populations to incorporate energy, which in turn depends on the life history of each species. This is how the impact propagates across the trophic web, differentially affecting species according to their role in the function and organization of an ecosystem. In this process, the response of the ecosystem in terms of its resilience implies that the probabilities of flows, represented by the entropy associated with the overhead, will change and redistribute according to the various properties associated with each node of the trophic network. In this condition, the final result is not the restoration of the ecosystem to a previous state, but the transition to an alternate state; that is, the same system reconfigured. At this point, we must remember that ecosystem resilience has two components: population contributions and ecosystem contributions, as a dynamic network. In our example, when all the exploited populations are functioning at their maximum production capacity, the population component will be altered, and the system will not have the opportunity to restore itself to the initial state.

In general, management measures for exploited populations are applied independently among fishery resources, even when the species belong to the same ecosystem. Some common management measures are temporal and spatial closures, minimum catch sizes, catch quotas, and resource access limitations, among others. If these measures are present in the conditions described in the previous example and we assume that the measures are adequate, recovery or maintenance of the biomass retained by the fishery per unit of time is expected, thus maintaining a stable and sustainable ecosystem and resources. However, if a contingency such as the one described in the same example occurs, management measures will no longer be sufficient to maintain the ecosystem's restorative capacity, and the ecosystem is no longer sustainable. These contingencies can be of natural origin, such as the effects of changes in climate patterns, or of human origin, such as fishing or alterations to habitats. Origin of contingencies is relevant from the moment at which they can be mitigated and/or controlled.

Holistic Attributes for Fisheries Management

If, in contrast to the previous perspective, ecosystem-based reference points are incorporated into the base criteria for management, the sustainability conditions could be maintained even in the presence of contingencies. For example, the definitions of resilience (Chap. 6) and of noxicline (Chap. 7) are based precisely on the role of entropy in an ecosystem, in the first case in relation to how a system uses entropy in the form of resilience; and in the second case in relation to the entropy generated by biomass extraction and the threshold limit to maintain the sustainability of the ecosystem. In this context, the usefulness of these concepts is based on not exceeding this threshold for the ecosystem to maintain its restoration capacity, while the system internally reorganizes itself. If the harvest rates are defined year to year within the threshold limits defined for the ecosystem and at the limit of the production capacity of exploited populations, sustainability can be maintained even under the influence of a continuous process of ecosystem reorganization.

In practice, management measures are easily transformed into tangible aspects that fishers and the sector in general understand well and, by nature, are, consequently, easy to implement; that is, they are manifested in control measures for extracted biomass or in limiting fishing effort. In all cases, both the control tools and the objective of such control are tangible aspects that fishers can measure, understand, and manage. On the other hand, when talking about ecosystem-based management, the conventional concept, "from the individual populations to the ecosystem," the management measures also turn out to be tangible elements, such as catch reductions, limited fishing areas, change of fishing gears, or its operation to protect undesirable impacts on the seabed, by either unwanted biomass removal or substrate removal; prevention or recovery of critical habitats. Also, contamination by biological or chemical waste is addressed through its collection or treatment, and species that are endangered, threatened, or in danger of extinction can be visually identified, all of which are elements that are familiar to fishers and managers and can be implemented.

Thus, for example, some holistic attributes of the ecosystem mentioned above, such as noxicline or resilience, are properties and concepts that are based on the probability of energy to be transfer between the ecosystem components, and although it is possible to recommend specific harvest rates for each resource, direct implementation is not easy to understand in practice because neither of the two attributes nor the information for their construction (entropy) represent tangible elements that can be easily and widely understood by the fishing sector. In this context, the use and application of the ecosystem holistic criteria through these types of indicators requires two essential elements: (1) that they be used in a complementary way with population indicators or reference points (conventional) and (2) that greater confidence be generated in the increasingly complex knowledge generated by science in benefit of the sustainable use of fish resources. In both the cases, the results lead, primarily, to the use of ecosystem-based biological reference points from a precautionary perspective, which, undoubtedly, will contribute not only to sustainable use but also to generating realistic expectations of growth and development of the fishing sector in the face of contingencies such as the effects of changes in climate patterns.

References

Arreguín-Sánchez, F. 1987. Present status of the red grouper fishery of the Campeche Bank. In *Proc. 37th. Annual Session Gulf and Caribbean Fisheries Institute, Martinique, 9–12 November 1985*, 498–509.

———. 2010. Cambio climático y el colpaso de la pesquería de camarón rosado (*Farfantepenaeus duorarum*) de la sonda de Campeche. In *Cambio Climático en México un Enfoque Costero-Marino*, 399–410. Campeche: Universidad Autónoma de Campeche.

———. 2012. The dynamics linking biological hierarchies, fish stocks and ecosystems: implications for fisheries management. In *Models of the ecological hierarchy: from molecules to the ecosphere*, ed. F. Jordán and S.E. Jørgensen, 501–516. Philadelphia, PA: Elsevier B.V.

———. 2019. Climate change and the rise of the octopus fishery in the Campeche Bank, Mexico. *Regional Studies in Marine Science* 32: 100852.

Arreguín-Sánchez, F., and T.J. Pitcher. 1999. Catchability estimates and their application to the red grouper (Epinephelus morio) fishery of the Campeche Bank, Mexico. *Fishery Bulletin*: 746–757.

Arreguín-Sánchez, F., J.C. Seijo-Gutiérrez, D. Fuentes-Castellanos, and M.J. Solís-Ramírez. 1987. *Estado del conocimiento de los recursos pesqueros de la plataforma continental de Yucatán y región adyacente*. México: Centro regional de Investigaciones Pesqueras de Yucalpetén, Instituto Nacional de Pesca. Documento Técnico No. 4: 41 p.

Arreguín-Sánchez, F., M.J. Zetina-Rejón, and M. Ramírez-Rodríguez. 2008. Exploring ecosystem-based harvesting strategies to recover the collapsed pink shrimp (*Farfantepenaeus duorarum*) fishery in the southern Gulf of Mexico. *Ecological Modelling* 214 (2–4): 83–94.

Arreguín-Sánchez, F., P. del Monte-Luna, and M.J. Zetina-Rejón. 2015. Climate change effects on aquatic ecosystems and the challenge for fishery management: pink shrimp of the southern gulf of Mexico. *Fisheries* 40 (1): 15–19.

Arreguín-Sánchez, F., P. del Monte Luna, M.J. Zetina-Rejón, A. Tripp-Valdez, M.O. Albáñez-Lucero, and T.M. Ruiz-Barreiro. 2017. Building an ecosystems-type fisheries management approach for the Campeche Bank, subarea in the Gulf of Mexico Large Marine Ecosystem. *Environmental Development* 22: 143–149.

Bessonov, N.M., A.A. Elizarov, and O. González. 1971. Rasgos principales de la formación de las condiciones oceanográficas. *Investigaciones Pesqueras Sovietico-Cubanas. Fascículo* 3: 14–32.

Brulé, T., C. Déniel, T. Colás-Marrufo, and M. Sánchez-Crespo. 1999. Red grouper reproduction in the southern Gulf of Mexico. *Transactions of the American Fisheries Society* 128 (3): 385–402.

Burgos-Rosas, R., and A. Pérez-Pérez. 2006. La pesquería de mero. In *Sustentabilidad y Pesca responsable en México. Evaluación y Manejo*, ed. F. Arreguín-Sánchez, L. Beléndez-Moreno, I. Méndez, R. Solana, and C. Rangel. México: SAGARPA, INAPESCA. 5644 p.

DOF. 2006. Carta Nacional Pesquera. *Diario Oficial de la Federación*. 25 Agosto 2006.

———. 2012. Carta Nacional Pesquera. *Diario Oficial de la Federación*. 13 Abril de 2012.

———. 2018. Carta Nacional Pesquera. *Diario Oficial de la Federación*. 11 Junio de 2018.

Echazabal-Salazar, O., E. Morales-Bojórquez, and F. Arreguín-Sánchez. 2021. Biomass dynamic model for multiple data series: An improved approach for the management of the red grouper (Epinephelus morio) fishery of the Campeche Bank, Mexico. *Regional Studies in Marine Science* 47: 101962.

García, C. 1980. Caracterización general del Banco de Campeche. *Revista Cubana de Investigaciones Pesqueras* 5 (2): 1.

Giménez-Hurtado, E. 2005. *Analisis de la pesca del mero Epinephelus morio (Serranidae: Pisces, Velenciennes 1928) en el Banco de Campeche. Tesis de Doctorado.* la Paz, Baja California Sur, México: ventro de Investigaciones Biológicas del noroeste. 93 p.

González, P., S. Zupanovich, and H. Ramis. 1974. Biología pesquera de la cherna americana del Banco de Campeche. *Revista Cubana de Investigaciones Pesqueras* 1: 107–111.

Pielke, R.A., Jr. 2004. What is climate change? *Energy & Environment* 15 (3): 515–520.

Solomon, S., M. Manning, M. Marquis, and D. Qin. 2007. *Climate change 2007-the physical science basis: working group I contribution to the fourth assessment report of the IPCC*. Vol. 4. Cambridge: Cambridge University Press.

Ulanowicz, R. 1986. *Growth and development: Ecosystems phenomenology*, 203. New York, NY: Springer.

———. 2009. The dual nature of ecosystem dynamics. *Ecological Modelling* 220: 1886–1892.

Walters, C., V. Christensen, and D. Pauly. 1997. Structuring dynamic models of exploited ecosystems from trophic mass-balance assessments. *Reviews in Fish Biology and Fisheries* 7 (2): 139–172.

Chapter 10
Ecosystem-Based Fisheries Management: Selected Examples

Ecosystem-Based Management

Ecosystem-based management is currently conceived as a decision-making process aimed at reducing the impact of the operation of a fishery on the ecosystem—in particular, the portion of the natural environment that is disturbed and on which fishing has a direct effect. Thus, for example, the incidental catch is identified as a problem, which is especially magnified when the nontarget species that are caught are subsequently discarded because they do not have a specific use or they lack economic value. Even when some species are used, there may be significant impacts due to both the large amount of biomass the species represent and the effects on the diversity of the species being killed. These types of incidental catch-related impacts occur in all fisheries; there are very few examples in the world of fisheries that only catch the target species, i.e., those that are strictly monospecific. As seen in Chap. 4, the production capacity of a population is determined by its life history, where part of it is associated with the intrinsic growth rate. Therefore, if several species with different life histories are simultaneously caught by the same fishing gear, each of them will respond with a different surplus production so that if the impact disappears, their restoration will not necessarily occur at the same rate, taking different times to reach the initial population size. Likewise, if the level of exploitation of the target species is high, it is likely that the surplus production of some species will not be sufficient to restore their biomass in a given unit of time. This poses at least a couple of major dilemmas: How much to fish of the target species so as not to affect other species in the ecosystem? How can bycatch be reduced? These problems are common to almost all fisheries although to different extents, with shrimp trawling probably being one of the most documented, where the bycatch volume is, on average, ten times larger than the catch of the target resource.

Another type of problem addressed under the current ecosystem approach has to do with the effects on endangered, threatened, and protected (ETP) species, for which society demands their conservation, as occurs with most marine mammals,

F. Arreguín-Sánchez, *Holistic Approach to Ecosystem-Based Fisheries Management*, https://doi.org/10.1007/978-3-030-96847-2_10

some birds and fishes, and other species. The effects of fishing on habitats are also considered, including water pollution as caused by fuel spills during fishing maneuvers; damage to the seafloor as caused by trawling; or in serious cases, the destruction of coral reefs by fishing gear.

Management oriented towards reducing or controlling bycatch assumes, in principle, a significant probability level of success since species retention is potentially directly "controllable" by the fishing operations. This is the case for the development of technologies that allow species to escape, such as turtle and fish excluder devices in trawls; the establishment of protected areas with specific objectives; specialized maneuvers performed to release species as occurs in tuna fishing with purse seine and dolphin release; control and/or collection of objects dumped into the sea; and, in general, the establishment of guidelines for the operation of fishing gear, among other strategies. However, all of these actions respond to the effect of a fishery on a part of the ecosystem; that is, in a perspective from the target resource to the ecosystem, or better defined, the portion of the ecosystem to which the fishery has access. The problem with this notion is that it does not recognize the ecosystem as a functional unit of nature, implicitly assuming that an ecosystem is the sum of its parts (populations), so that if one part deteriorates, controlling fishing will permit restoration, assuming this as sufficient.

As shown in Fig. 1.4, populations and ecosystems are, biologically speaking, two hierarchical levels with different complexities and dynamics in nature. The dynamic performance of a population could be measured through its survival; given that to persist and maintain itself, it requires a capacity for renewal. In an ecosystem, however, persistence depends on the maintenance of its organization through its topology (connectivity through trophic relationships) and its functioning (energy flows between species occurring mainly through predator-prey relationships). Thus, the performance and maintenance of an ecosystem depends on the maintenance and restoration capacity of these connections and energy exchanges. In the "from the individual populations to the ecosystem" approach, what is truly valued is the maintenance of populations, whether they are fishing targets, of interest for conservation or incidentally caught; the restoration of habitats these species inhabit; and fishing control, through management decisions, is ultimately carried out to limit the mortality or physical effects on habitats exerted by fishing. In this context, the concept of ecosystem sustainability truly corresponds to the sustainability of the populations.

The ecosystem approach seen from the previous perspective does not consider the effects of fishing, including the effects of the applied control measures, on ecosystem organization and consequently, on ecosystem topology and functioning. In this context, a key question arises: How much of a resource's biomass should remain in the sea, after fishing, so as not to affect ecosystem organization and functioning? The answer to this question requires knowing ecosystem attributes related to its functioning and organization and from there, identifying the feasible harvest rate to apply in a "from the ecosystem to the individual populations" approach. Of course, this approach also includes aspects related to bycatch, conservation of ETP species, habitat quality, etc., from the perspective of ecosystem organization and functioning.

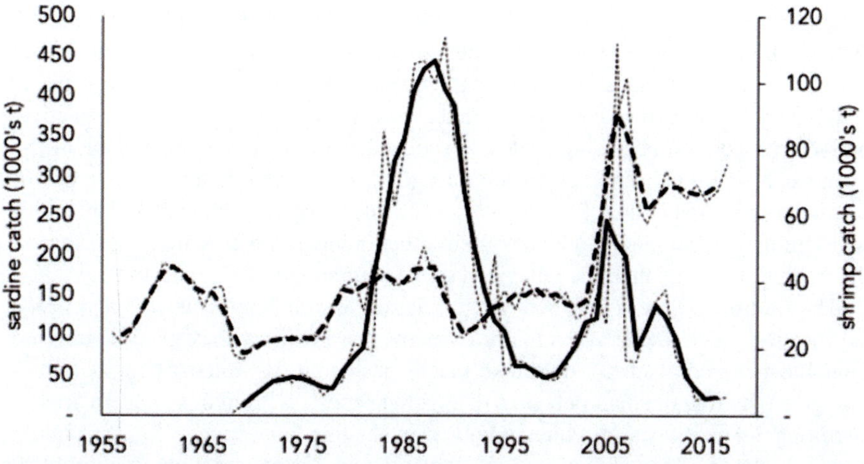

Fig. 10.1 Historical time series of catches for the two main fisheries in the Gulf of California, sardine (solid line), and shrimp (dashed line). Lines represent a 3-year moving average and dotted thin lines the observed data

This chapter presents examples of management problems of some fishery resources that inhabit different types of ecosystems, and the objective is to illustrate the contribution to ecosystem sustainability of the approach called "from the ecosystem to the individual populations."

Monterrey Sardine (*Sardinops sagax*) in the Central Gulf of California

The central Gulf of California is one of the most productive fishing regions of Mexico. Historically, the most important fisheries by volume were shrimp, tuna, and sardine, which were all industrial fisheries, whereas mojarra, skipjack tuna, crab, and squid were the species with the highest historical catches obtained by small-scale coastal fisheries, although species such as snapper, grouper, lobster, and clams were of great importance due to their market value. The tuna fishing is recorded in this region, but it is truly an oceanic fishery, which of course includes the Gulf of California, since the fleet is deployed throughout the Exclusive Economic Zone of Mexico in the Pacific Ocean. The rest of the species are caught mainly on the continental shelf or in its vicinity. Sardine and shrimp resources are both at low trophic levels and are considered commercial species that in turn are the base of food webs, that is, forage species. Both are considered resources exploited at their maximum biological production capacity (DOF 2018). The two species have shown fluctuations in catch, which are clearly larger for sardine than for shrimp (Fig. 10.1). In both the cases, fluctuations in catch have been related to climate patterns affecting population abundance (Lluch-Belda et al. 1999; Lluch-Cota et al. 2010), with the benthic environment being more stable, since it is associated with detritus, than the pelagic environment, which is associated with phytoplankton abundance. In the particular case of sardine, under the current regulatory perspective, periods of decrease in fishing yields, associated with decreased abundance of the resource, have often been interpreted as a sign of overfishing. In this context, one of the major concerns is that if fishing causes this decrease in sardine abundance, this will seriously affect the overall ecosystem functioning as this resource is the base of the trophic web on which the pelagic system is supported.

The sardine fishery is certified by the Marine Stewardship Council,[1] and in the certification process, one of the major concerns was precisely that, given a decrease in abundance, overfishing would take place that would cause ecosystem deterioration given the role of small pelagic fish as forage species. From this concern arose a condition for certification which involved evaluating how much biomass should remain in the sea after fishing to avoid damaging ecosystem functioning, a problem that was addressed by Arreguín-Sánchez et al. (2021).

This concern reached the public and, given a decrease in the abundance and catches of Monterey sardine in 2017 and 2018, a public initiative was started proposing the establishment of a biosphere reserve area in the Sea of Cortés, with the main objective being the conservation of both marine mammals and aquatic diversity, but largely supported by the assumption of sardine overfishing and the effect this would have on the ecosystem because this species is at the base of the trophic web.

The answer to the initial question of how much biomass should remain in the sea after fishing to avoid affecting the ecosystem was addressed by estimating the threshold harvest rate based on holistic indicators using the concepts of noxicline (Arreguín-Sánchez et al. 2022a) and resilience (Chap. 6); the former as a measure of the limit of disturbance to ecosystem organization (ecosystem entropy gain); and the latter as a limit to avoid affecting the resilient capacity of the system.

Figure 10.2 shows the results of the estimates of the threshold harvest rate (HR) for the sardine fishery according to the proposed holistic indicators, with a HR = 0.38 in both the cases. According to these results, the threshold catch or biomass retention of Monterey sardine is 38% of the available biomass and is similar both to preserve ecosystem organization and functioning and to maintain its resilient capacity.

The usefulness of these concepts lies in the fact that the estimation of the biomass extraction threshold is independent of the magnitude of the available biomass since it is a ratio between catch (*C*) and biomass (*B*): HR = *C*/*B*; that is, the magnitude of the catch will be determined by the available biomass at any given time. In the particular case of the Monterey sardine, the harvest rate coincides between both estimators and is consistent with an optimal harvest rate of HR = 0.35 in the late 1980 and

[1] https://fisheries.msc.org/en/fisheries/small-pelagics-fishery-in-sonora-gulf-of-california/.

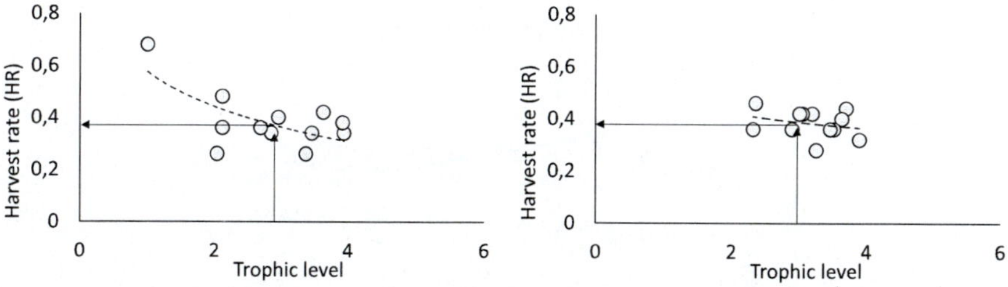

Fig. 10.2 Limit-harvest rates for the sardine fishery in the central Gulf of California derived from holistic ecosystem attributes: on the left, the noxicline (limit of exploitation associated with the increase in entropy in the ecosystem, Arreguín-Sánchez et al. 2022a); and resilience, on the right. For both, the harvest rate was HR = 0.38 which means that 62% of the available biomass must remain at the sea to avoid an undesirable impact of fishing on the ecosystem

HR = 0.29 by the late 1990 (Martínez-Aguilar et al. 2009), both from stock assessment models.

Mangrove Community in Magdalena Bay, South Baja California

The literature has reported the importance of mangrove ecosystems in supporting fisheries, as they serve as fishing areas, but mainly as breeding areas, where they provide habitat, protection against predation and, directly or indirectly, food for many species. In particular, mangrove leaves and wood, through their incorporation into detritus as a consequence of their degradation, are of great importance in the food chains that support fisheries (i.e., Hutchison et al. 2014; Manson et al. 2005). Several studies have also addressed the role of mangroves with respect to ecosystem services, and particularly, to the economic effect of fisheries, either by economically assessing the fishery production supported by mangroves or by assessing the magnitude of catches in terms of biomass (i.e., Barbier 2000; Aburto-Oropeza et al. 2008; Yulianto et al. 2016; Fitri et al. 2018). The ecosystem services provided by mangroves have been documented as important, recognizing the need for their conservation, but failing that, for their restoration when they have been disturbed (Botero and Salzwedel 1999; Duke et al. 2007; Malik et al. 2015; López-Portillo et al. 2017; Rafique 2018).

In general, research on mangroves associated with fishing has focused on the services that the mangrove system provides to the mangrove system itself and to adjacent systems. While the dependence of many coastal systems on mangrove systems is recognized, there are no clear and explicit reports on the effects of mangroves on the organization and function of adjacent aquatic ecosystems, where species of economic interest inhabit.

In this context, Chavez-Rosales (2006) developed a model in which the mangrove system is associated with the adjacent estuarine system, valuing especially the

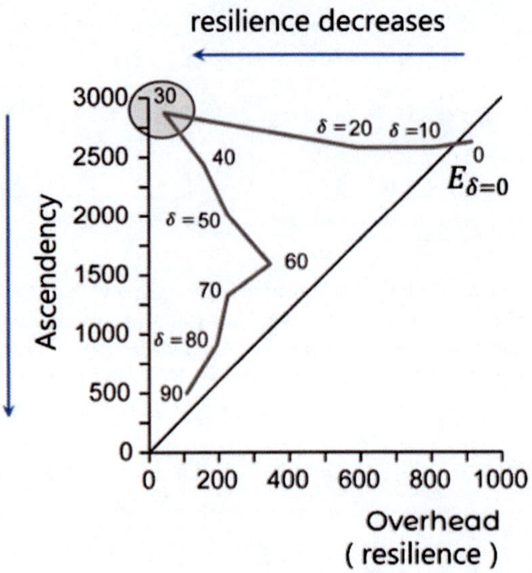

Fig. 10.3 Effect of the loss of mangrove cover in the adjacent aquatic system for Magdalena Bay in the western coast of the Peninsula of Baja California. Ascendency represents ecosystem organization; the overhead is an index for resilience. Numbers represents the percent of mangrove cover lost. At low levels of coverage loss, less than 30%, system can recover, even when resilience is diminishing. After such level, ecosystem cannot recover and organization is deteriorated with the increase in loss of mangrove coverage

contribution of mangroves to the aquatic food web through the input of organic matter associated with detritus. From this model, a simulation experiment was conducted in which the mangrove cover was gradually reduced and the attributes of the aquatic ecosystem were measured, including the region adjacent to the mangrove. The attributes measured were ascendency, as an index of organization, and overhead, as an indicator of ecosystem resilience (Ulanowicz 1986, 2009). The results were expressed in terms of ecosystem health, following Ulanowicz (1992), Costanza and Mageau (1999), and Costanza (2012). According to Fig. 10.3, the state of disturbance (loss) of mangrove cover is expressed by the symbol δ, and the associated number reflects the percentage of disturbance. With each level of disturbance, the aquatic system is disturbed. As shown in Fig. 10.3, as mangrove cover is lost, ecosystem organization, represented by ascendency, is maintained but shows loss of resilience capacity, represented by overhead. This decrease continues as the mangrove loss increases, reaching a threshold, which in this case was a 30% loss in mangrove cover, where resilience is minimal. By increasing the loss of mangrove cover, the aquatic ecosystem loses organization, always maintaining a low level of resilience.

The mangrove area associated with Magdalena Bay encompasses an area of approximately 218.5 km^2, and the adjacent estuarine system covers an area of 958.9 km^2, with recorded landings by coastal fisheries of approximately 2500 tons/year (Aburto-Oropeza et al. 2008). According to the simulation experiment, the aquatic ecosystem adjacent to the mangrove would lose its resilient capacity and organization if a mangrove area of 65.5 km^2 is lost. This type of data is important in terms of restoration policies, given that the Magdalena Bay region is exposed, year after year, to the impact of storms and hurricanes. Although this phenomenon is worrisome, the mangrove swamp in Magdalena Bay seems to have increased its cover by approximately 20% over the last two decades, and this may be associated with an increase in mean sea level (López-Medellín et al. 2011). Even under these conditions, the simulation exercise illustrates a procedure of interest to define a threshold reference level in mangrove cover loss as a precautionary strategy.

Red Grouper (*Epinephelus morio*) from the Campeche Bank

Campeche Bank comprises the northern and western continental shelf of the Yucatan Peninsula towards the southern Gulf of Mexico. It is a region that, due to its oceanographic characteristics, is clearly different from the rest of the marine waters of the Gulf of Mexico. It comprises a broad continental shelf, approximately between 18° and 24° north latitude and 87° and 93° west longitude (Fig. 10.4). Biological production strongly depends on phytoplankton biomass since there are no surface rivers providing a significant amount of nutrients, and terrestrial nutrient input is very limited and only has an influence near the coast. Marine currents are of low intensity; therefore, turbulence processes on the seabed also do not contribute significantly to sustaining global production. The peninsula is surrounded by eddies along the outer limits of the continental shelf that function as a barrier to water exchange with the rest of the Gulf of Mexico (Arreguín-Sánchez et al. 2019). This is one of the marine regions in Mexico where the effects of climate change are most clearly manifested. Del Monte-Luna et al. (2015) reported that a low-frequency, harmonic component (67 years) explains just over 60% of the sea-surface temperature on Campeche Bank, while Arreguín-Sánchez et al. (2017a) presented evidence of regime shift in the area, including a reduction in global primary production.

The red grouper fishery (*Epinephelus morio*) has historically been the second most important fishery in the Gulf of Mexico after shrimp. The resource is exploited by three fleets that sequentially catch the resource with varying degrees of overlap. There is a small-scale fleet that operates nearshore, typically at depths less than 30 m; a Mexican semi-industrial fleet operating below 20 m; and a Cuban fleet operating below 35 m; the latter two moves towards deeper waters, reaching an overlap of approximately 80%.

The current status of the red grouper fishery is considered deteriorated as it is overexploited (DOF 2012, 2018). In particular, there has been a decreasing trend in catches starting in the early 1990s (Fig. 10.5), with the current yields of the species being approximately 40% of the highest historical record (early 1970s and 1990s). It is actually a multispecies fishery where the proportion of red grouper in the catch has decreased from 95% in the 1970s to near 50% in the current era. According to the National Fisheries Charter (DOF 2012, 2018), the current status of the red grouper

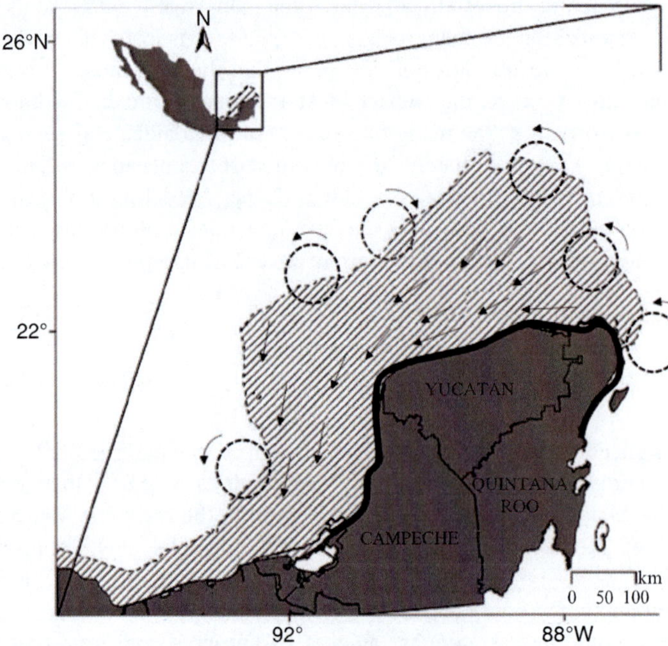

Fig. 10.4 Diagram showing the hydrodynamics, of the sea in the Campeche Bank. The dashed area represents the continental shelf; the arrows the main direction of currents, which are of low intensity; circles represent eddies around the continental shelf. All eddied run contrary to clockwise with exception to one on the north central region. This behavior limits the interaction with the central Gulf of Mexico

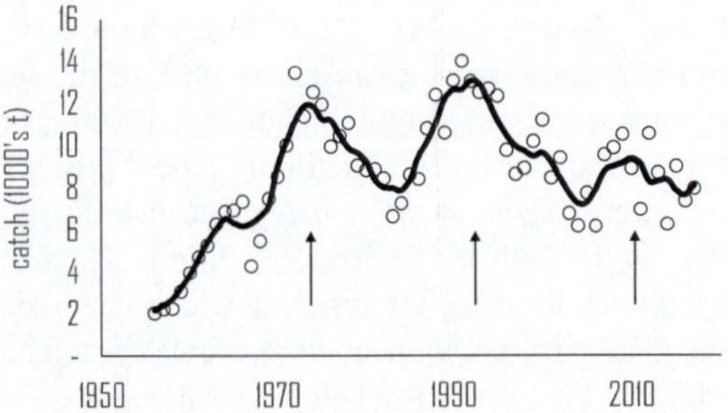

Fig. 10.5 Time series of catch of the red grouper (Epinephelus morio) in the Campeche Bank. Arrows highlight decadal fluctuations. Despite such fluctuations, the third peak is about 70% of the previous and recent trends yields are near 50% of the yields at early 1990s, and 40% of the maximum sustainable yield. (See Echazabal-Salazar et al. 2021)

fishery is clearly deteriorated, and it is estimated that the current biomass is below the biomass that maximizes surplus production. Given that the red grouper population aggregates during the reproductive season, fishing is suspended during this time; besides, a minimum landing size is applied.

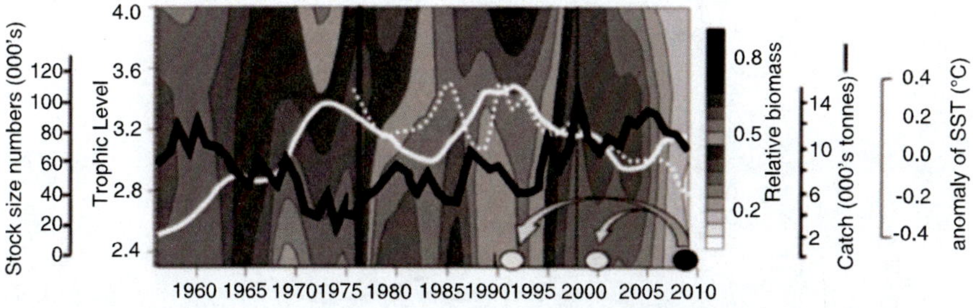

Fig. 10.6 Evolution of the ecosystem of the north continental shelf of Yucatan in the southern Gulf of Mexico during six decades. Different grays indicate changes in the relative biomass, where the clearest corresponds to lower amount of biomass. Black solid line represents the anomaly of the SST; white solid and dotted white lines represent red grouper catches and stock size, respectively. Arrows starting from the recent year to backwards represent the objectives of current management strategies. Here observation is made that the structure of the ecosystem, and with it its function, was different in the times to which it is intended to restore the population of red grouper. This leads to the assumption that the management measures will not have the expected efficiency, since the ecosystem in the near future no necessarily will have the same condition. (The figure was derived from simulation experiments through the Ecosim model (Walters et al. 1997), incorporating time series of fishing and environmental forcing)

Arreguín-Sánchez et al. (1996) observed decadal cycles in the size of the grouper population, suggesting the effect of environmental patterns, while Arreguín-Sánchez et al. (2017a, b) analyzed various resources and found evidence of the effects of climate change on several species from Campeche Bank, including red grouper. Figure 10.6 shows the inverse trends in sea-surface temperature and red grouper population size. The red grouper is a long-lived species whose age of first maturity occurs at 4 years; therefore, the result of the current breeding stock generation will be observed as part of the breeding stock catchable the following year. It has been documented that red groupers require relatively low temperatures for maturation and reproduction (Bessonov et al. 1971; García 1980), and these waters are found in the winter in the central eastern region of the Yucatan continental shelf, where the fish aggregate for reproduction. Elevated temperatures tend to inhibit maturation, resulting in lower reproductive success. When the temperature rises consistently, year after year, the cumulative effect on the population manifests as a decrease in population size.

According to the National Fisheries Charter (DOF 2012, 2018), red grouper resources are below the threshold reference point, which corresponds to an average biomass of 52,000 tons in the last two decades. If the deterioration was caused only by fishing, recovery could be expected by reducing the harvest rate; however, despite the limitations imposed on fishing, the resource did not respond as expected. As shown in Fig. 10.6, the ecosystem has been reconfigured as induced by environmental forcing; in particular for red grouper, abundance and catches have decreased in synchrony with the increase in temperature.

An ecosystem that is continually being reconfigured as a result of environmental forcing, and in particular the grouper population that adapts to this circumstance, it is

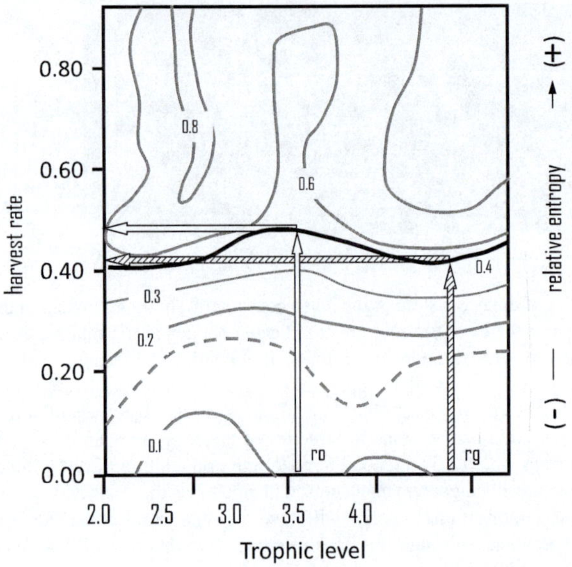

Fig. 10.7 Ecosystem biological reference point, EBRP, for the red grouper (*Epinephelus morio*) and the red octopus (*Octopus maya*) of the north continental shelf of Yucatan ecosystem defined by the harvest rates. Species are identified by their trophic level. For the red grouper $HR_{EBRP\_RG} = 0.42$, and for the red octopus $HR_{EBRP\_OC} = 0.48$. Black bold line represents the noxicline for the whole ecosystem (Arreguín-Sánchez et al. 2022a). Such harvest rates mean that each fishing season 58% and 52% of the available biomasses must remain at the sea, after fishing, respectively, to avoid impact on the organization and function of the ecosystem

possible to guide the exploitation strategy, suggesting in this case the one that defines the biomass limit as a criterion. To be retained by fishing that allows maintaining the organization and functioning of the ecosystem, a level represented by the noxycline (Arreguín-Sánchez et al. 2022a). For red grouper (Fig. 10.7), the recommended harvest rate is HR = 0.42; that is, no more than 42% of the available biomass can be extracted in any given year.

The expectation is that the combination of the harvest rate estimated by holistic criteria (noxicline) and measures to avoid growth and recruitment overfishing can guide the sustainable exploitation of red grouper in a dynamic way, according to the interactions between fishing, climate, and interspecific relationships.

Pink Shrimp (*Penaeus duorarum*) from Campeche Bank

The pink shrimp (*Penaeus duorarum*) from Campeche Bank has historically been the most important fishery resource for the Mexican region of the Gulf of Mexico. Fishing is carried out in the western region of the Yucatan Peninsula, known as Campeche Sound. For approximately two decades and until the early 1970s, the yields of this species reached approximately 20,000 tons/year. Since the second half of the 1970s, coinciding with a regime shift, the abundance of pink shrimp has been decreasing consistently as a result of a consistently decreasing recruitment rate. Currently, the abundance of this resource is estimated to be between 5% and 7% of the abundance in the 1970s. In 1995, the collapse of this fishery was recognized, and it was initially assumed that it was due to the effects of overfishing, particularly on growth overfishing and to negative effects on the habitat by the fishery and the beginning of the oil industry in the 1980s (Arreguın-Sánchez 2001). This condition, however, does not coincide with the historical trend of pink shrimp abundance or with changes in fishing effort. Arreguín Sánchez et al. (2008) showed that recruitment had been decreasing continuously since the mid-1970s, with a strong inverse association with the temperature trend and a direct association with the also decreasing primary production.

Until 1979, American, Cuban, and Mexican fleets operated in this fishery, with more than 1000 vessels, with only the Mexican fleet remaining after 1980. Between 1982 and 1984, the ownership of the shrimp fleet was transferred from the private sector to the cooperative sector, and part of the fleet ceased activity. Between these two events, the effective fishing effort was reduced by approximately 50%; however, no significant response by pink shrimp was observed. As mentioned above, changes in climatic patterns and in the particular conditions of Campeche Bank (i.e., Fig. 10.4) have induced a process of continuous ecosystem reconfiguration through the adaptation and response of populations to environmental forcing. In the case of pink shrimp in Campeche Sound, the ecosystem has historically shown variations with a trend associated with climatic forcing, such as temperature, as shown in Fig. 10.8, where the highest relative biomass was observed in the mid-1970s, decreasing towards the current era, with approximately decadal fluctuations. Likewise, a decrease in the global carrying capacity of the ecosystem was observed from the beginning of the warming period. From the viewpoint of the fishery, the declared objective is to recover the stock to the abundance levels observed in the mid-1990s and, eventually, to the abundance levels of the 1980s (DOF 2012, 2018). It seems clear that this management objective has little probability of being achieved given that the state of the current ecosystem is very different from that existing in those two eras, with the global production capacity of the population being approximately 30% of what it was in the 1980s.

Given that the ecosystem has been undergoing continuous reconfiguration for several decades, the harvest rate corresponding to the noxicline (Arreguín-Sánchez et al. 2022a) was estimated as the threshold exploitation level that allows sustainable fishing and a sustainable ecosystem (Fig. 10.9), estimated at an HR = 0.33; that is, maximum extraction of 33% of the available biomass each year, which allows, in addition to the restoration of the population biomass extracted by fishing, the maintenance of ecosystem organization and functioning.

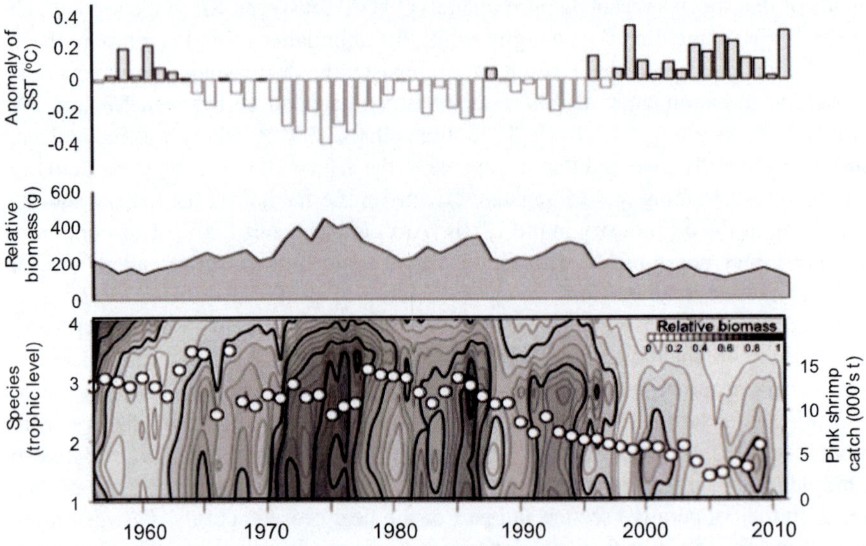

Fig. 10.8 Evolution of the ecosystem of the Campeche Sound on the southern Gulf of Mexico during six decades. Different grays in the low panel indicates changes in the relative biomass, where the clearest corresponds to lower amount of biomass. White dots indicate pink shrimp catches. Gray area on the middle panel represents the carrying capacity, and the bar diagram on top represents the anomaly of SST. Note that at the end of the cooling period, the ecosystem production reaches its maximum production, decreasing later following the warming tendency. Also note the decadal fluctuations but always following the large-term trend. Shrimp population responded to such climate pattern by diminishing its abundance and catches as consequence. The production panel along the time was simulated through the Ecosim model (Walters et al. 1997), incorporating time series of fishing and environmental forcing (adapted from Arreguín-Sánchez et al. 2015)

Red Octopus (*Octopus maya*) from Campeche Bank

In contrast to other fisheries of Campeche Bank, the red octopus (*Octopus maya*) was favored by the warming trend starting in the 1980s. According to Solis-Ramírez and Arreguín-Sánchez (1984) and Arreguín-Sánchez et al. (1987), this resource was fully exploited in the 1980s, a condition that was maintained until the mid-1990s (Arreguín-Sánchez et al. 2000), with annual yields close to 8000 tons. In recent decades, consistent increases in fishery yields have been observed. Arreguín-Sánchez et al. (2019) showed that from the beginning of the warming period in the 1980s, the abundance of red octopus has increased and, as a consequence, its fishery has also increased. This increase in abundance is associated with the increase in the robustness of individuals and reproductive efficiency. According to the trend shown, the fishery reached another period of relative stability in the early 2000s, with several years of maximum yields of 15,000–20,000 tons/year (Fig. 10.10).

With regard to the evolution of the ecosystem, Arreguín-Sánchez et al. (2017a, 2019) noted a significant reduction in the abundance of predators, particularly red grouper, which is contrary to that in the 1980s, when it was reported that during

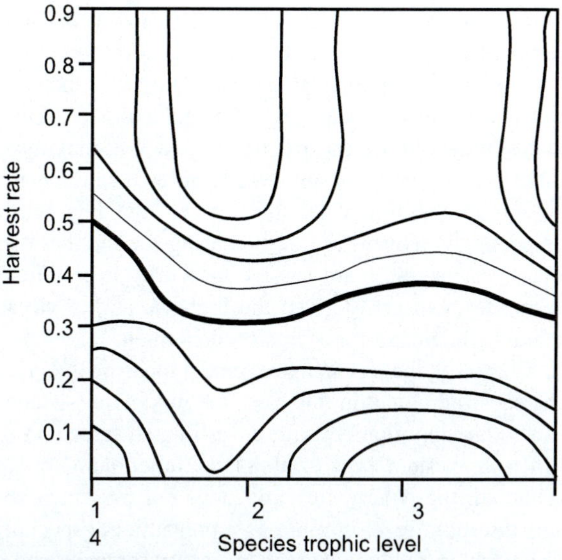

Fig. 10.9 Ecosystem biological reference point, EBRP, for the pink shrimp (Penaeus duorarum) of the Campeche Bank ecosystem defined by the harvest rate. Species is identified by its trophic level. For the pink shrimp, the reference level is given by $HR_{EBRP\_RG} = 0.34$, which means 66% of the available biomass must remain at the sea to avoid impact on the organization and function of the ecosystem. The black bold line represents the noxicline for the whole ecosystem (Arreguín-Sánchez et al. 2022a)

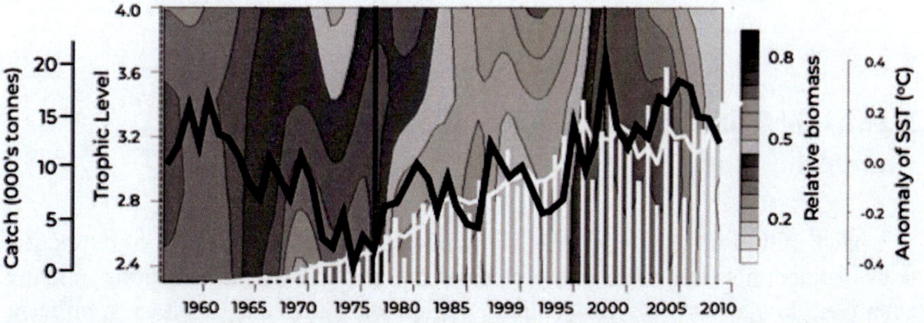

Fig. 10.10 Evolution of the ecosystem of Atasta coast, on the southern Campeche Bank along six decades. Different grays indicate changes in the relative biomass, where the clearest corresponds to lower amount of biomass. White bars and line indicate red octopus catches. The black bold line is the anomaly of SST. Changes in the ecosystem configuration are observed with the beginning of the warming period, as well as an increase in annual catches. The production panel along the time was simulated through the Ecosim model (Walters et al. 1997), incorporating time series of fishing and environmental forcing

higher red octopus abundance in the coastal zone (August to December), the presence of red octopus in the stomach contents of top predators was frequent and common, and red grouper made up approximately 50% of the abundance of predators. The reduction in the abundance of predators, particularly red grouper (Fig. 10.6), decreased predation pressure and, under favorable conditions, red octopus abundance has increased.

The changes in ecosystem configuration over the last six decades, its relationship with the warming trend starting in the 1980s, and the effect on the abundance of the red octopus population, as illustrated by trends in catches, is shown in Fig. 10.10. In terms of management of the red octopus resource, it is evident that its response to changes in environmental patterns has been increased production, which consequently has increased fishing effort to achieve full use of the resource. The current status of the red octopus fishery is defined as maximum sustainable use (DOF 2012, 2018), with population-related measures having been put in place, such as a closed season and minimum landing size to avoid growth and recruitment overfishing, respectively, as well as not increasing the number of fishing permits issued. However, because these are passive measures, it is difficult to contend with short-term responses, especially given the high variability in catches in recent years and the continued demand for access by fishermen.

Currently, there is no management measure linked to the role of the red octopus in the ecosystem, and in this case, the noxicline (Arreguín-Sánchez et al. 2022a) provides this opportunity, with an estimated threshold harvest rate for fishing not to affect ecosystem organization and functioning of an HR = 0.48% (Fig. 10.7). Although the fishery currently does not present problems, it is important to keep this threshold level in mind as a precaution, especially when dealing with a population that is highly vulnerable to an environment where the environmental forcing is essentially cyclical.

Hawksbill Turtle (*Eretmochelys imbricata*) from Campeche Bank

The hawksbill turtle (*Eretmochelys imbricata*) is a species at the risk of extinction that is subject not only to special protection but also to restoration actions in many countries. In the case of the southern region of Campeche Bank, a significant decrease in the frequency of nesting has been detected. Del Monte-Luna et al. (2012) found a significant relationship between the change in nesting intensity and the environmental patterns in the region, particularly temperature, and the phase change that occurred in the mid-1990s. It is evident that the ecosystem as a whole has been reconfigured as a consequence of the adaptation of populations. In this context, Fig. 10.11 shows this synchrony of changes in the biomass structure of the populations in the ecosystem (and therefore in ecosystem organization) with environmental changes, including the particular case of the hawksbill turtle, manifested through nesting intensity.

In terms of conservation-oriented management, lack of knowledge of the effects of changes of climate patterns on species and ecosystems can be critical. When this information is unknown, it is typical to assign these changes (in this case, in nesting intensity) to local causes such as impacts on habitats or fishing, or undocumented mortality, to which the obvious answer would be to identify the related activities that have a real potential or assumed interaction with the turtle population and take the appropriate actions to mitigate the observed effects. In the case of the hawksbill

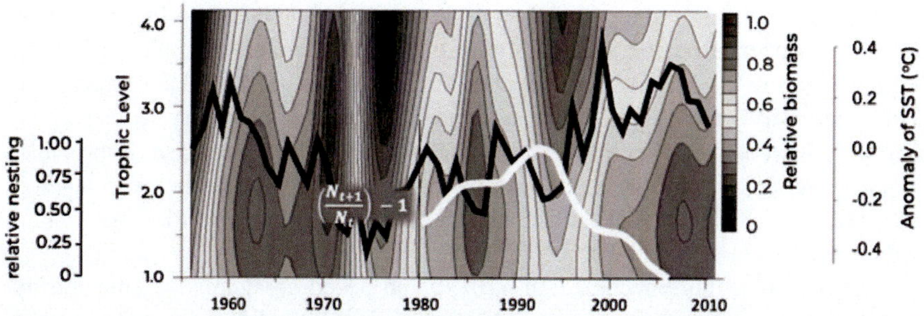

Fig. 10.11 Evolution of the ecosystem of the western coast of the Peninsula of Yucatan along six decades. Different grays indicate changes in the relative biomass, where the clearest corresponds to lower amount of biomass. The black bold line represents the anomaly in SST, and the white line represents a proxy of the stock abundance given by the number of nests on beach of the hawksbill turtle (*Eretmochelis imbricata*). The turtle (nesting) stock strongly decline with an apparent strong increase in the warming tendency which also appears to affect ecosystem structure (and as consequently its organization and function). The beginning of the declining period coincides with the change of phase, from negative to positive, of the anomaly of SST (see Del Monte-Luna et al. 2012). The production panel along the time was simulated through the Ecosim model (Walters et al. 1997), incorporating time series of fishing and environmental forcing

turtle from southern Campeche Bank, what seems to be happening is the manifestation of the effect of a natural change on the ecosystem, where conservation actions could be guided by adaptability strategies aimed at preserving the best possible conditions in the usual nesting habitats to allow turtles to have adequate nesting opportunities. Regarding hawksbill turtle populations, it is known that this species is highly migratory, and although this has not been documented, it is likely that the species has switched to alternative nesting areas throughout its distribution range.

Top Predators of the Gulf of California

One aspect that is often of concern is the assignment of responsibility for ecosystem impacts to fishing. Beyond the question of whether the negative effects of fishing on some ecosystems are reasonable or not, or are real or not, it should not be ignored that regulations and management measures have social consequences for families and communities that depend on fishing. Obviously, the establishment of regulations implies limits, but the effectiveness of regulations must be based on the adequate identification of the sources of variation in resource abundance, including the interaction effects due to changes in such abundances. Proper identification of the sources of variation will have direct consequences on the expected efficiency of regulations and management measures.

In the Gulf of California, Sala et al. (2004) argued that coastal fisheries are unsustainable and degrade the ecosystem and that management measures must be re-evaluated to counteract an inefficient fishery. Their conclusions are based on an analysis of the changes in top predators in the ecosystem in the period from 1950 to 2000, with information obtained mainly from interviews with fishermen. Their main arguments were as follows: (1) a decreasing trend in catches and an increase in fishing intensity; (2) a change in the structure of catches of predatory species in the 1970s–1980s compared with that in the 1990s–2000s; (3) a decrease in the mean trophic level of catches; (4) changes in fishing areas; (5) changes in mean landing size; and (6) a change in community structure. These authors concluded that fishing is the causative agent for these changes.

Recently, Arreguín-Sánchez et al. (2017b) proposed an alternative situation by considering the contribution of the effect of changes in climatic patterns on the abundance of resources and consequently on the identified status of fisheries in combination with fishing. Figure 10.12 shows that the changes referred to by Sala et al. (2004) in the structure of catches for the two periods mentioned in (2) in the previous paragraph correspond to two different climatic conditions: colder temperatures in the 1970s–1980s and warmer temperatures in recent decades. This difference in climatic regimes seems clear, including, with respect to the greater diversity of species in catches in the decades with a higher temperature, a process recognized in the literature as "tropicalization" (Pauly 1998; Vergés et al. 2014; Cheung et al. 2013; Horta e Costa et al. 2014; Liang et al. 2018).

Based on the argument of the effect of SST as driver of variation with a tendency of abundances of fish resources, and a tendency towards a tropicalization effect, it is possible to explain some of the aspects identified by Sala et al. (2004). The decrease in the catches of some top predators seems to have been associated with the regime shift in the early 1980s, clearly identifying this effect in the fish communities already established in the new regime in the 1990s. The structure of the catches between periods (Fig. 10.12), the decrease in the average trophic level, and the change in the structure of the communities all could be evidence of tropicalization; even changes in fishing areas seem to be due to changes in fishing strategies of fishermen to new ecosystem conditions. Changes in the mean size of the catch may also have been the result of the same process of tropicalization, of increased fishing intensity, or both.

One of the problems with regulation and management, given the lack of information, is to separate with a good level of confidence the specific contribution of the different sources of variation in the abundance of exploited populations. In other words, only fishing can be controlled; and if it is possible to characterize the sources of variation as those due to climate, then there will be a greater variation explained and it will be possible to better define, with a greater degree of confidence, specific measures with an acceptable potential for success. In contrast, if the natural variation is assumed to be due to fishing, the risk of failure in the expected results of the applied management measures will be high. In turn, it is clear that if fishing continues unchanged and abundance decreases, the first diagnosis will obviously be overfishing; or, if the actual abundance decreases but the yields per boat are maintained, overfishing will necessarily result. The relative advantage of estimating threshold reference points from holistic criteria is that the weight of the estimation is focused on maintaining the organization and function of the ecosystem, even when it is continuously reconfigured. The use of the noxicline concept provides the

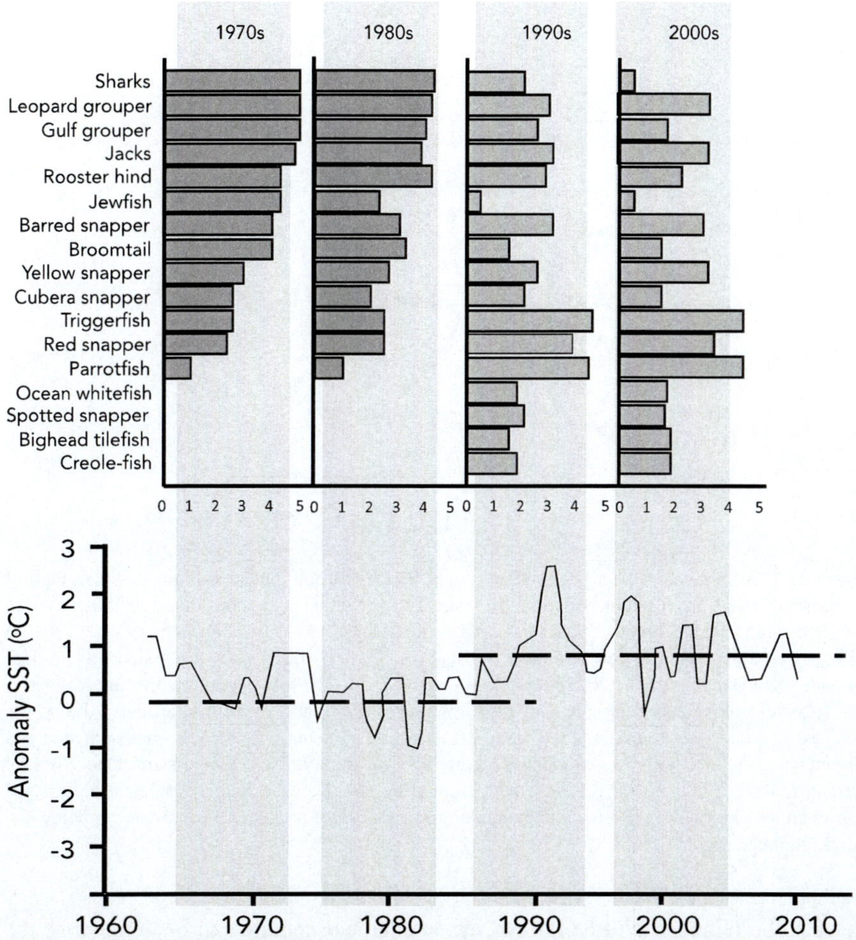

Fig. 10.12 Changes on predators' communities observed over decades in the Gulf of California and initially assumed as impact of overfishing (Sala et al. 2004), and an alternative or complementary hypothesis suggesting that observed changes has been influences by changes in climate patterns. The anomaly was computed for the period of 1880–2010. (Adapted from Arreguín-Sánchez et al. 2017b)

possibility of determining, through the recommended harvest rate, the proportion of biomass that must remain in the sea to maintain ecosystem organization and functioning, an estimate that depends on the available biomass at any given time. Thus, for example, for the top predators of the Gulf of California (trophic levels above 3.5, which includes all species shown in Fig. 10.12), a harvest rate of HR = 0.30 (Fig. 10.13a) is suggested, which indicates that up to a maximum of 30% of the biomass at the sea could be retained by fishing in a given year. Thus, if the available biomass is estimated at the beginning of the year or fishing season, the maximum

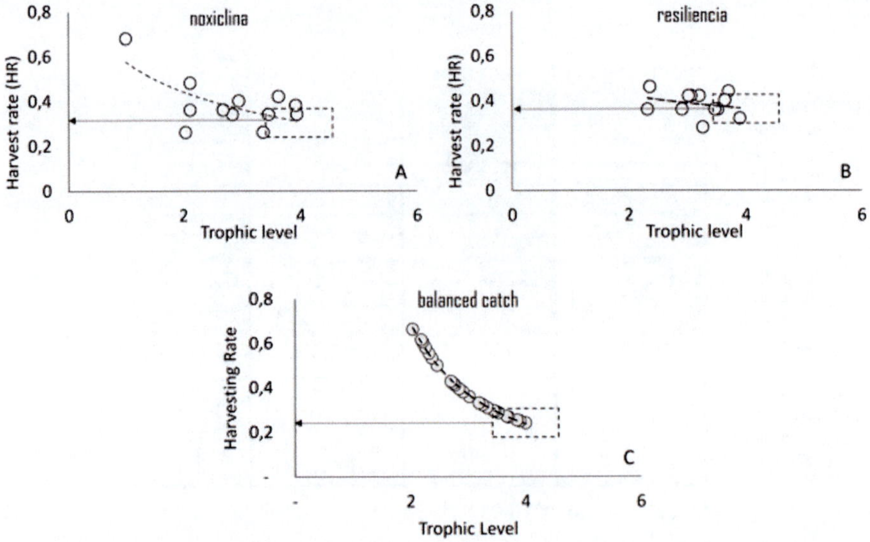

Fig. 10.13 Approach to harvest rates identified as EBRPs for top predators of the northern Gulf of California, based on different holistic indicators. Top predators were considered as those species with trophic level higher than 3.5. For noxicline (Arreguín-Sánchez et al. 2022a) $HR_{EBRP\_TP\_nox} = 0.30$; for resilience $HR_{EBRP\_TP\_res} = 0.36$; and for balanced production (Arreguín-Sánchez et al. 2022b) $HR_{EBRP\_TP\_BP} = 0.24$. The first two, noxicline and resilience, are relatively close and correspond to limit reference points and mean that around 70% of the biomass at the sea must remain at the sea to avoid impact of fishing on the ecosystem function and organization; while the balanced catch suggests 25% of the available biomass must be taken by fishing (and 75% remains at the sea), which could be considered as target reference point since represent the exploitation level where the maximum capacity of production is used and is below the limit reference points

catch to be removed will be known. An analogous scenario can be defined for the resilient capacity of the ecosystem, where the harvest rate would be HR = 0.37 (Fig. 10.13b). Evidently we are talking here about the effects of fishing but, if other human activities negatively affects stock abundance, the threshold reference points must be defined considering the total loss of biomass.

The condition posed by Sala et al. (2004), although it contrasts with the previous argument, allows for considering the importance of ecosystem-based management. The theory of population dynamics, and in particular the maximum surplus production, implies maintaining the population in a state equivalent to 50% of its maximum biomass, corresponding to the carrying capacity of the system. This would be equivalent, in terms of the harvest rate, to a value of HR = 0.50. If the resources in the Gulf of California are regulated based on the concept of maximization of surplus production, the harvest rate would be higher than that defined by both the noxicline (Arreguín-Sánchez et al. 2022a) and the production capacity derived from the concept of balanced production (Arreguín-Sánchez et al. 2022b) (Figs. 7.8 and 10.13c), both resulted in an HR = 0.30. This difference in terms of resource management implies that, in the first case, there are likely to be risky situations of overexploitation of top predators given that the maximization of production implies an application of 65% more fishing intensity than recommended by the balanced capture and noxicline criteria.

As in all cases reviewed in the previous sections, harvest rates are derived from holistic criteria. When fishery regulations are defined and management measures are designed, it should be remembered that populations are only one of the biological hierarchies of nature, composed of individuals and that they interact with other populations, and they configurate ecosystems. Populations and ecosystems are biological entities with their own dynamics and, in this sense, are interdependent; being the first, in system analytical terms, nested in the second. To maintain sustainable dynamic ecosystems, management measures must consider attributes of both biological hierarchies, especially those characterizing their interdependence, in addition to their own particular attributes. In the previous examples, emphasis is placed on harvest rates derived from holistic criteria, but in each case, it will be necessary to consider essential population aspects, such as those related to avoiding growth and recruitment overfishing and the processes that determine recruitment success and fishing efficiency, among others.

References

Aburto-Oropeza, O., E. Ezcurra, G. Danemann, V. Valdez, J. Murray, and E. Sala. 2008. Mangroves in the Gulf of California increase fishery yields. *Proceedings of the National Academy of Sciences* 105 (30): 10456–10459.

Arreguín-Sánchez F., P. del Monte-Luna, M.J. Zetina-Rejón, L.O. Duarte, D. Lercari, M. Riofrío-Lazo, M.O. Albáñez-Lucero, Danilo Calliari, T. M. Ruiz-Barrreriro, R. Vögler. 2022a. Towards an ecosystem holistic approach to estimate limit harvesting rates.

Arreguín-Sánchez F., M.J. Zetina-Rejón, P. del Monte-Luna, M.O. Albáñez-Lucero, L.O. Duarte, D. Lercari, D. Calliari, R. Vogler. 2022b. Reference harvest rates estimated from trophic level and production capacity of exploited stocks.

Arreguín Sánchez, F., M. Ramírez Rodríguez, M.J. Zetina Rejón, and V.H. Cruz Escalona. 2008. Natural hazards, stock depletion, and stock management in the southern Gulf of Mexico pink shrimp fishery. In *American Fisheries Society Symposium*.

Arreguın-Sánchez, F. 2001. Towards the management of fisheries in the context of the ecosystem: the case of Mexico. *European Commission Fisheries Cooperation Bulletin* 14: 7–9.

Arreguín-Sánchez, F., J.C. Seijo, D. Fuentes-Castellanos, and M.J. Solíz-Ramírez. 1987. *Estado del conocimiento d elos recursos pesqueros de la plataforma cntinental de Yucatán y región adyacente. Contribuciones de Investigación pesquera*. México: Centro regional de Investigación pesquera, Instituto Nacional de Pesca. Documento Técnico No. 4, 41 p.

Arreguín-Sánchez, F., M. Contreras, V. Moreno, R. Burgos, and R. Valdés. 1996. Population dynamics and stock assessment of red grouper (Epinephelus morio) fishery on Campeche Bank, Mexico. In *Biology, fisheries and culture of tropical groupers and snappers: ICLARM Conference Proceedings*, 202–217.

Arreguín-Sánchez, F., M.J. Solís-Ramírez, and M.E. González De La Rosa. 2000. Population dynamics and stock assessment for Octopus maya (Cephalopoda: Octopodidae) fishery in the Campeche Bank, Gulf of Mexico. *Revista de Biología Tropical* 48 (2–3): 323–331.

Arreguín-Sánchez, F., P. Del Monte-Luna, and M.J. Zetina-Rejón. 2015. Climate change effects on aquatic ecosystems and the challenge for fishery management: pink shrimp of the southern gulf of Mexico. *Fisheries* 40 (1): 15–19.

Arreguín-Sánchez, F., P. del Monte Luna, M.J. Zetina-Rejón, A. Tripp-Valdez, M.O. Albáñez-Lucero, and T.M. Ruiz-Barreiro. 2017a. Building an ecosystems-type fisheries management approach for the Campeche Bank, subarea in the Gulf of Mexico Large Marine Ecosystem. *Environmental Development* 22: 143–149.

Arreguín-Sánchez, F., P. del Monte-Luna, M.J. Zetina-Rejón, and M.O. Albáñez-Lucero. 2017b. The Gulf of California large marine ecosystem: fisheries and other natural resources. *Environmental Development* 22: 71–77.

Arreguín-Sánchez, F., M.O. Albañez-Lucero, P. Del Monte-Luna, and M.J. Zetina-Rejón. 2019. Fishery resource management challenges facing climate change. In *Mexican aquatic environments*, 181–194. Cham: Springer.

Arreguín-Sánchez, F., P. del Monte-Luna, M.O. Albañez-Lucero, M.J. Zetina-Rejón, A. Tripp-Quezada, T.M. Ruiz-Barreiro, and J.C. Hernández-Padilla. 2021. How much biomass must remain in the sea after fishing to preserve ecosystem functioning? The case of the sardine fishery in the gulf of California, Mexico. In *Marine coastal ecosystems modelling and conservation*, ed. M. Ortiz and F. Jordán, 143–161pp. Cham: Springer Nature.

Barbier, E.B. 2000. Valuing the environment as input: review of applications to mangrove-fishery linkages. *Ecological Economics* 35 (1): 47–61.

Bessonov, N.M., A.A. Elizarov, and O. González. 1971. Rasgos principales de la formación de las condiciones oceanográficas. *Investigaciones Pesqueras Sovietico-Cubanas. Fascículo* 3: 14–32.

Botero, L., and H. Salzwedel. 1999. Rehabilitation of the Ciénaga Grande de Santa Marta, a mangrove-estuarine system in the Caribbean coast of Colombia. *Ocean and Coastal Management* 42 (2–4): 243–256.

Chavez-Rosales, S. 2006. *El papel de los manglares en la producción de las comunidades acuáticas de Bahía Magdalena, B.C.S. Tesis Doctor en Ciencias*. La Paz, Baja California Sur: Instituto Politécnico Nacional, Centro Interdisciplinario ee Ciencias Marinas. 143 p.

Cheung, W.W., R. Watson, and D. Pauly. 2013. Signature of ocean warming in global fisheries catch. *Nature* 497 (7449): 365–368.

Costanza, R. 2012. Ecosystem health and ecological engineering. *Ecological Engineering* 45: 24–29.

Costanza, R., and M. Mageau. 1999. What is a healthy ecosystem? *Aquatic Ecology* 33 (1): 105–115.

DOF. 2012. Carta Nacional Pesquera. *Diario Oficial de la Federación*. 13 abril de 2012. https://www.gob.mx/inapesca/acciones-y-programas/carta-nacional-pesquera-51204.

———. 2018. Carta Nacional Pesquera. *Diario Oficial de la Federación*. 11 junio de 2018. https://www.gob.mx/inapesca/acciones-y-programas/carta-nacional-pesquera-51204.

Duke, N.C., J.O. Meynecke, S. Dittmann, A.M. Ellison, K. Anger, U. Berger, S. Canicci, K. Diele, K.C. Ewel, C.D. Field, N. Koedam, S.Y. Lee, C. Marchand, I. Nordahus, and F. Dahdouh-Guebas. 2007. A world without mangroves? *Science* 317 (5834): 41–42.

Echazabal-Salazar, O., E. Morales-Bojórquez, and F. Arreguín-Sánchez. 2021. Biomass dynamic model for multiple data series: an improved approach for the management of the red grouper (*Epinephelus morio*) fishery of the Campeche Bank, Mexico. *Regional Studies in Marine Science* 47: 101962.

Fitri, A., M. Basyuni, R. Wati, N. Sulistiyono, B. Slamet, Z.A. Harahap, et al. 2018. Management of mangrove ecosystems for increasing fisheries production in Lubuk Kertang village, North Sumatra, Indonesia. *AACL Bioflux*: 1252–1264.

García, C. 1980. Caracterización general del Banco de Campeche. *Revista Cubana de Investigaciones Pesqueras* 5 (2): 1–10.

Horta e Costa, B., J. Assis, G. Franco, K. Erzini, M. Henriques, E.J. Gonçalves, and J.E. Caselle. 2014. Tropicalization of fish assemblages in temperate biogeographic transition zones. *Marine Ecology Progress Series* 504: 241–252.

Hutchison, J., M. Spalding, and P. zu Ermgassen. 2014. *The role of mangroves in fisheries enhancement*. Vol. 54. Arlington, VA; Wageningen: The Nature Conservancy and Wetlands International.

Liang, C., W. Xian, and D. Pauly. 2018. Impacts of ocean warming on China's fisheries catches: an application of "mean temperature of the catch" concept. *Frontiers in Marine Science* 5: 26.

Lluch-Belda, D., S. Lluch-Cota, D. Lluch-Cota, and S. Hernández-Vázquez. 1999. La variabilidad oceánica interanual y su impacto en las pesquerías. *Revista de la Sociedad Mexicana de Historia Natural* 49: 219–227.

Lluch-Cota, S.E., A. Parés-Sierra, V.O. Magaña-Rueda, F. Arreguín-Sánchez, G. Bazzino, H. Herrera-Cervantes, and D. Lluch-Belda. 2010. Changing climate in the Gulf of California. *Progress in Oceanography* 87: 114–136.

López-Medellín, X., E. Ezcurra, C. González-Abraham, J. Hak, L.S. Santiago, and J.O. Sickman. 2011. Oceanographic anomalies and sea-level rise drive mangroves inland in the Pacific coast of Mexico. *Journal of Vegetation Science* 22 (1): 143–151.

López-Portillo, J., R.R. Lewis, P. Saenger, A. Rovai, N. Koedam, F.D. Dahdouh-Guebas, C. Agraz-Hernández, and V.H. Rivera-Monroy. 2017. Mangrove forest restoration and rehabilitation. In *Mangrove ecosystems: a global biogeographic perspective*, 301–345. Cham: Springer.

Malik, A., R. Fensholt, and O. Mertz. 2015. Mangrove exploitation effects on biodiversity and ecosystem services. *Biodiversity and Conservation* 24 (14): 3543–3557.

Manson, F.J., N.R. Loneragan, G.A. Skilleter, and S.R. Phinn. 2005. An evaluation of the evidence for linkages between mangroves and fisheries: a synthesis of the literature and identification of research directions. In *Oceanography and marine biology*, 493–524.

Martínez-Aguilar, S., J.A. de Anda-Montañez, F. Arreguín-Sánchez, and M.Á. Cisneros-Mata. 2009. Constant harvest rate for the Pacific sardine (Sardinops caeruleus) fishery in the Gulf of California based on catchability-at-length estimations. *Fisheries Research* 99 (2): 74–82.

del Monte-Luna, P., V. Guzmán-Hernández, E.A. Cuevas, F. Arreguín-Sánchez, and D. Lluch-Belda. 2012. Effect of North Atlantic climate variability on hawksbill turtles in the Southern Gulf of Mexico. *Journal of Experimental Marine Biology and Ecology* 412: 103–109.

del Monte-Luna, P., H. Villalobos, and F. Arreguín-Sánchez. 2015. Variability of sea surface temperature in the southwestern Gulf of Mexico. *Continental Shelf Research* 102: 73–79.

Pauly, D. 1998. Beyond our original horizons: the tropicalization of Beverton and Holt. *Reviews in Fish Biology and Fisheries* 8 (3): 307–334.

Rafique, M. 2018. A review on the status, ecological importance, vulnerabilities, and conservation strategies for the Mangrove ecosystems of Pakistan. *Pakistan Journal of Botany* 50 (4): 1645–1659.

Sala, E., O. Aburto-Oropeza, M. Reza, G. Paredes, and L.G. López-Lemus. 2004. Fishing down coastal food webs in the Gulf of California. *Fisheries* 29 (3): 19–25.

Solis-Ramírez, M.J., and F. Arreguín-Sánchez. 1984. Analysis of octopus (Octopus maya) fisheries in the Campeche Bank. In *Mem. IX MEXUS-GULF Meeting, Cancun, Quintara Roo, 15-–16 November 1984, México*, 31–35.

Ulanowicz, R. 1986. *Growth and development: ecosystems phenomenology*, 203. New York, NY: Springer.

Ulanowicz, R.E. 1992. Ecosystem health and trophic flow networks. In *Ecosystem health: new goals for environmental management*, ed. R.C.B.G. Norton, M. Faber, and D. Rapport, 190–206.

Ulanowicz, R. 2009. The dual nature of ecosystem dynamics. *Ecological Modelling* 220: 1886–1892.

Vergés, A., P.D. Steinberg, M.E. Hay, A.G. Poore, A.H. Campbell, E. Ballesteros, K.L. Heck, D.J. Booth, M.A. Coleman, D.A. Feary, W. Figueira, T. Langlois, E.M. Marzinelli, T. Mizerek, P.J. Mumby, Y. Nakamura, M. Roughan, E. van Sebille, A.S. Gupta, D.A. Smale, F. Tomas, T. Wernberg, and S.K. Wilson. 2014. The tropicalization of temperate marine ecosystems: climate-mediated changes in herbivory and community phase shifts. *Proceedings of the Royal Society B: Biological Sciences* 281 (1789): 20140846.

Walters, C., V. Christensen, and D. Pauly. 1997. Structuring dynamic models of exploited ecosystems from trophic mass-balance assessments. *Reviews in Fish Biology and Fisheries* 7 (2): 139–172.

Yulianto, G., K. Soewardi, and L. Adrianto. 2016. The role of mangrove in support of coastal fisheries in Indramayu Regency, West Java, Indonesia. *Aquaculture, Aquarium, Conservation & Legislation* 9 (5): 1020–1029.

Chapter 11
Conclusions

Population-Based Fisheries Management

It has been mentioned in various forums that extractive fishing in the world has reached its maximum production capacity; this is shown by the information from FAO on the state of world fisheries, where the extractive fishing of marine waters has been sustained, in the last 3–4 decades, in around 90 million tons/year, and those of fresh water in around 10 million annually. In this way, the challenge of fishing management, in the first instance, is with the global sustainability of fishing yields and what this means in natural and socioeconomic terms, not only that, but also with the commitment for development, which means, among other things, reducing both poverty and food insecurity.

The challenge not only refers to the destiny or expected achievements of the management, but also how to achieve it. Conventionally, and because natural conditions allowed it, it has been assumed that fishing resources have a growth limit defined as the system's carrying capacity and that this varies according to their life history. According to this, stability has been assumed in natural systems, this being the baseline for decision-making since practically any measure is based on the idea of reaching or maintaining a certain biomass production (capture), or to restore production, to certain desired levels, but always, taking the baseline defined by said load capacity as the basis of the calculation.

Of course, the management under this context is not easy, because practically no fishery is monospecific as is assumed in the population models used as a scientific basis to advise decision-making. Fishing for multiple target species is complicated because different life histories of stocks imply differences in production capacity. This generates differences in yields, be they in volume of biomass or in the value of the catches, or even in the social benefits derived from each type of resource. In general, the management, in these cases, will be aimed at avoiding the risks of overfishing of any of the target resources, and within this context, it will be the fishermen who will define the specific orientation of their activities. In this sense, the

F. Arreguín-Sánchez, *Holistic Approach to Ecosystem-Based Fisheries Management*, https://doi.org/10.1007/978-3-030-96847-2_11

population model(s) applied to the resources will indicate the permissible conditions. Note that under this concept, neither biological interdependencies nor the impact of climate patterns are considered at any time, assuming stability. In this way, it is expected that the management measures respond to the control of fishing mortality.

With regard to bycatch, including discards, the greatest concern refers to negative impacts on biodiversity; in two senses, the massive mortality of species, as occurs in the shrimp and finfish trawl fisheries in tropical regions, where bycatch averages 10 tons for each ton of the target species, being mostly species that are discarded. Another aspect refers to the effect that removal of large volumes of biomass has on ecosystems. It is presumed that it has negative effects on their organization and operation, but there is little quantitative information that provides evidence of the damage and its consequences. There is also concern about the mortality caused on species considered endangered, threatened, and protected (ETP), on which, in some cases, there are specific control measures are taken. The same type of situation occurs in many small-scale fisheries although bycatch volumes are often unrecorded and bycatch is assumed to be smaller than those from industrial trawl fisheries. On the other hand, an additional aspect refers to the effect of fishing on the marine habitats, especially the seabeds with which the fishing gears interact directly. The effects on the seabed can be very varied, depending on the ecosystem, the fishing gear and the intensity of the fishing operations. In these cases, the control measures have to do with the interruption of the impact on the seabeds, generally by way of prohibitions to carry out fishing in critical habitats, such as coral reefs or coastal areas that serve as nurseries.

In the last 2–3 decades, the concept of ecosystem-based fisheries management has been mentioned with increasing frequency. Although this concept contemplates many of the aspects mentioned for conventional population-based approaches, it recognizes the importance of interdependencies between species; particularly those represented by the predator–prey relationships. The energy exchanges among species in an ecosystem can be characterized by specific processes, of course, the trophic relationships, but also processes such as migrations, exports, or system dissipation of energy because of respiration. In general, the energy exchanges that represent the attributes of the living ecosystem are represented by the food web. Even if the simplest example is assumed, without migration and without export, most of the energy flows between species. In this way, the dynamics of ecosystems is usually represented by the network attributes characterized and expressed by the dynamics of trophic relationships.

The size of an ecosystem is determined by the carrying capacity of the environment, and in this context by the availability of nutrients and basic elements that allow sustaining life in a given space and time, and by the ability to be used by living organisms. In general, as a consequence of the evolutionary process, it is assumed that an established ecosystem has grown and developed using the elements of the environment with a certain efficiency, which, according to its dynamics, has resulted in the development and establishment of the different species and communities, and a balance of the thermodynamic properties of the trophic system. Consequently, if the system is not disturbed, it is assumed that it is structured, organized, and functioning in a state of thermodynamic balance; that is, a state in which the biomass of populations may vary, but global energy exchanges remain stable. This principle of thermodynamic balance has been useful for the trophic modeling process of ecosystems.

As mentioned in the initial chapters, individuals, populations, and ecosystems are, each one, a functional unit in the nature, with their own and independent dynamics and performance that intrinsically characterize them, but that also interact with each other and are self-dependent. The characteristics and performance of an ecosystem are usually measured by the attributes of the food web, in such a way that the knowledge of its structure and the basic understanding of its dynamics is essential to quantify the effects of disturbances such as the loss of biomass caused by the fishing. The loss of biomass will be reflected immediately through the alteration of energy flows in the food web. Predators will have less food from a particular prey (if this is the target species for fishing) and to balance their requirement they will consume more of other prey and these in turn will respond in the same way, and the effect spreads throughout the system, with different intensities. The magnitude of each interaction will be defined by different aspects such as food preferences, effects of density-dependence, competition, distribution, etc., of each of the species. If fishing target is a top predator, the effect spreads towards the preys, which will increase its biomass by being less consumed, and these in turn will consume a little more of their prey, and so on. The particular effects of the decrease in biomass of a population will be seen in this way, reflected in the entire ecosystem, and the impact can be quantified through performance indicators of the trophic web. The contrast of the attributes of the net, with and without the effect of biomass extraction, or the comparison at two different moments with different extraction levels, will provide base information for use in fisheries management, provided that these performance indicators can be transformed in harvest rates, and finally in fishing mortality.

From the point of view of the manager or the fisherman, the management practice, once the control rules have been defined, is applied by identifying the practical elements, whether they are catch limits, or limits to the application of fishing effort. In both the cases, for the management measures to be easily adopted, it is important that the decision-maker, the fisherman, the recipient, and the executor identify in the most tangible way the possible origin, reason, and destiny of the matter under ordinance. As this information is better understood, the response for its implementation will be better accepted and put into practice; and consequently, more effective. In the practice of conventional management, measures are easily identified because they refer to tangible elements; either biomass limits through catches or limits on the fishing mortality which is easily translated into fishing effort. In both the cases, the measures obey, for example, to maintain a determined spawning biomass, to guarantee a certain remnant of biomass in the sea, to allow the survival of recruits to grow and reproduce, or to limit the mortality of the ages with the highest reproductive efficiency, among others; and all of them are tangible elements for the manager as well as for the fisherman. If the reasons, the objects under control, and benefit are understood and perceived in a tangible way, the application of the ordinance will be easily implemented and with a high probability of success.

Ecosystem-Based Fisheries Management

In ecosystem-based management, there are aspects that are tangible for the fisherman; for example, those concerning mortality due to fishing that affects biodiversity relating to ETP species. However, on many occasions when talking about biodiversity, as a concept, it is hard to put measures into practice because, although the concept is understood, the consequences of its impact are not tangibly and nor clearly observable. For example, answering a fisherman the question about what biodiversity is for, or why it is important for the functioning of the ecosystem, is not an easy task; of course, regarding the effect and benefit or not for fishing. Another example refers to the resilience of populations which, however, can be perceived because it is, in part, associated with the population growth rate and this, in turn, can be tangible through the success of recruitment and profit in abundance. This stage is the one defined in Chap. 2 as ecosystem-based management under a strategy "from individual stocks to ecosystems."

Populations and ecosystems are two different biological hierarchies, with also different performance attributes. Populations provide reasonably tangible arguments for management effects that can be measured and observed; this means that the public perception of management and its application can even be followed and its degree of success measured. On the other hand, populations are the components of ecosystems, where the attributes of the system are defined mainly through the interdependencies between these populations, and more specifically through trophic relationships represented as energy exchanges. In this context, the performance attributes of the ecosystem do not correspond to those for populations but to their organization, functioning and implicitly to their structure, all of them related to the dynamics of the trophic web. In this way, when an impact occurs on any of the populations, not only this is affected, but also those species with which it interacts directly and in turn with which the latter interact, and so on; that is, the impact signal spreads via the food web impacting the whole ecosystem. In this sense, the energy needed for the ecosystem response comes from the two hierarchies of biological complexity; populations, through surplus production (see Chap. 4), and energy from adjustments in the food web (see Chap. 6). An inherent characteristic of food webs refers to the fact that responses to a disturbance are not linear in nature with respect to interactions; that is, the responses depend to a great extent on the particular characteristics of the species, for example, the response seen through the interaction between two species of high trophic level, regarding two of low trophic level, is different; or, between one of a high trophic level and another of a low trophic level, where the response will depend on the characteristics of the life histories of each species, on the quantity and quality of trophic interconnections, on the strategies that allow them to take the available food resources, of the differences in the responses to density-dependent effects, of their interactions with the habitat, etc.; hence it is easy to understand the nonlinearity in the responses when comparing them between species, and even when comparing the role of the same species in two neighboring ecosystems. The direct consequence of this is that any ecosystem-based management strategy must be specific and/or adapted to each particular system, especially when trying to maintain its sustainability.

The sustainability of an ecosystem should not be understood as the stability of the production capacity. The performance of the food web is measured by the energy flows between its components, in such a way that the sustainability of the ecosystem should be measured through its performance in terms of its capacity of organization and its functioning; that is, from attributes of the food web. If, as Ulanowicz (1986, 2009) suggests, the organization of the ecosystem is measured through Ascedency (or some indicator related to organization), and this, according to Eq. 3.1 (Chap. 3), is defined by the combination and arrangement of the flows of the whole system, it is then possible to find different states of the ecosystem for the same index of organization or condition, which essentially means different combinations of abundances of populations and energy exchanges that result in the same indicator value; that is, more than one state of organization and functioning of the ecosystem can represent a same sustainable state. In the event of a disturbance, if the biomass is recovered by the performance of the population itself (e.g., the surplus production or production capacity), a relative stability of the ecosystem organization can also be expected since the system will invest relatively little energy in said restoration. However, if the disturbance exceeds the population's restoration capacity, the ecosystem compensatory processes come into play with greater intensity through the food web (see Chap. 6) and, if the disturbance is of great magnitude, changes in the ecosystem will manifest as changes in its organization and function, impacting the ecosystem resilience capacity. In accordance with the above, and in terms of fishing activity, sustainable management based on the ecosystem should not only consider the sustainability of populations, but also of the ecosystem, in such a way that both population and ecosystem indicators should be considered. In this sense, the ecosystem holistic indicators, based on trophic network attributes, also suggest ecosystem reference points in terms of harvest rates to be considered for fisheries management purposes, that is, a strategy "from the ecosystem to the individual populations" (see Chap. 2).

A problem of the practice of management under a strategy "from the ecosystem to the individual populations" is that the reference points or performance indicators of the trophic network, when estimated from the energy flows dynamics, are not tangible; that is, they are hard to understand by the fishing sector. Suppose, as shown in several of the examples in Chap. 10, that a harvest rate of HR = 0.40 is defined (i.e., only 40% of the biomass available in the sea can be extracted) under the argument of that by extracting more than that proportion of biomass, the organization of the ecosystem is affected or that its resilience is affected. The concepts of organization and resilience of the ecosystem are attributes difficult to understand and intangible for the daily practice of management; that is, even though it has been shown in previous chapters (see Chap. 10) that it is possible to estimate specific harvest rates for independent stocks from performance indicators or holistic ecosystem reference points, their direct use in practice may seem complicated as they are not tangible for the fishing sector. In contrast to this perspective of complexity towards the practice of management, the justification for the consideration of holistic reference points of the ecosystem lies precisely in the need for management under conditions of dynamic sustainability of the ecosystem and its resources. The current management challenge for sustainable fishing has to do with the continuous changes in population dynamics and also continuous ecosystem' reconfiguration over time.

In this context, it should be remembered that at the population level, sustainability can be measured through the survival of the population; that is, indicators reflected in terms of biomass or number of individuals; while at the ecosystem level, sustainability will be defined by network attributes in terms of their organization (or changes in entropy) and resilience, expressed as energy flows. Both the criteria are of great relevance given that in the responses to disturbances, both the population and the ecosystem participate with intrinsic and interdependent compensatory biological processes in accordance with their biological hierarchy. In conditions where the ecosystem, affected by both fishing and climate patterns, is continually reconfigured, its sustainability, seen as the maintenance of the organization and natural functioning, is interpreted as a dynamic event where the ecosystem can show alternative states without ceasing to do a sustainable use of its resources. In this context, management should be based on complementary information from both the population and the ecosystem. On the one hand, there are biological reference points at the population level, which, according to the biomass available year after year, will offer part of the information on the considerations necessary for its sustainability. On the other hand, the ecosystem reference points provide information on the role of resources in the ecosystem, allowing the identification of the critical disturbance threshold of the system because of the biomass loss. This threshold corresponds to the level from which the organization and function of the ecosystem are irreversibly affected (see the concept of noxycline and resilience, Chaps. 5 and 6). Since both the reference points reflect on harvest rates, the global strategy could consider the ecosystem-based reference level as a precautionary orientation, while the population reference point could determine specific actions in the short term.

Of course, contrasting or even conflicting situations will be found. For example, if the harvest rate suggested by the biological reference point of the ecosystem for a species such as tropical shrimp is HR = 0.40, it will mean that no more than 40% of the biomass available in the sea should be extracted; however, the population biological reference point corresponding to the maximum sustainable yield suggests a harvest rate of HR = 0.50. Of course, the latter will be preferred by the sector and the argument could be supported by the fact that it is a strategist population "*r*", with a high population growth rate, high fertility, and high population resilience. Another case is the minor pelagic species, with population attributes also corresponding to a strategic species "*r*" and which have been said to be the basis of the trophic chain in the pelagic environment. An analogous situation, although it is not a properly holistic performance indicator, is the shark harvest rate suggested through the balanced production concept (Chap. 4, Fig. 4.5) of HR = 0.26, which contrasts with the reference point HR = 0.5 corresponding to the maximum sustainable yield level. In this type of situation, from the ecosystem perspective, the complementarity of both population and ecosystem indicators should be considered, and consequently the suggestion will be, in general, to choose the management strategy that, in a precautionary way, allows meeting the objectives of sustainability of the use of fish resource. However, it will be necessary to generate trust and understanding in the fishing sector to accept holistic indicators for resource management, as part of the approach "from the ecosystem to the individual populations" (see chap. 2).

In the definition of management strategies, the effects of changes in climate patterns on populations and ecosystems must necessarily be considered. On the one hand, because they explain part of the changes in the abundance of resources, and their identification allows the effects of fishing to be determined with greater precision; and as a consequence, the expected results of the management. On the other hand, as mentioned above, trending changes in climate patterns will also generate continuous changes in the organization of the ecosystem since they will be subject to continuous reconfiguration. In both the cases, the adjustments to the management measures must be carried out also continuously, in such a way as to make it possible to foresee the necessary adaptations year after year; that is, an adaptability strategy must be incorporated into the management systems.

References

Ulanowicz, R. 1986. *Growth and development: ecosystems phenomenology*, 203. New York, NY: Springer.

———. 2009. The dual nature of ecosystem dynamics. *Ecological Modelling* 220: 1886–1892.

Index

F. Arreguín-Sánchez, *Holistic Approach to Ecosystem-Based Fisheries Management*, https://doi.org/10.1007/978-3-030-96847-2